Microbial Resolution

Proximities

Experiments in Nearness

David Cecchetto and Arielle Saiber, Series Editors

Published in association with the Society for Literature, Science, and the Arts

---

Microbial Resolution: Visualization and Security in the War against Emerging Microbes
Gloria Chan-Sook Kim

MICROBIAL RESOLUTION

Visualization and Security in the War against Emerging Microbes

Gloria Chan-Sook Kim

Proximities

University of Minnesota Press
Minneapolis
London

The University of Minnesota Press gratefully acknowledges the generous assistance provided for the publication of this book by the University of California–Riverside College of Humanities, Arts, and Social Sciences Book Subvention Fund.

Portions of chapter 1 are adapted from "Pathogenic Nation-Making: Media Ecologies and American Nationhood under the Shadow of Viral Emergence," *Configurations* 24, no. 4 (2016): 441–70, https://doi.org/10.1353/con.2016.0029. Portions of chapter 3 are adapted from "Microbial Scale and the Undoing of Vision," *ASAP Journal* 6, no. 1 (2021): 59–67, https://doi.org/10.1353/asa.2021.0005.

Published by the University of Minnesota Press
111 Third Avenue South, Suite 290
Minneapolis, MN 55401-2520
http://www.upress.umn.edu

ISBN 978-1-5179-1169-0 (hc)
ISBN 978-1-5179-1170-6 (pb)

A Cataloging-in-Publication record for this book is available from the Library of Congress.

Printed in the United States of America on acid-free paper

UMP BmB 2024

Contents

Preface

An indeterminate boundary dissolves into the atmosphere. A form emerges but resists definition. The image on the cover of this book oscillates between materialization and dematerialization, emergence and disappearance. It represents the visual trace of a pigmented bacteria (*Serratia marcescens*) propelling itself around a plate of agar, captured on a digital camera and saved onto a computer as a downsampled low-resolution image. To render the image at the quality required for printing, it was later upsampled and manipulated to a point that stressed the bit-rate capacities of its compressed information. In addition to capturing microbial logics and logistics, then, this image indexes the technological interventions that fill in the gaps of its informational compression and distention. The resulting blur might elicit an automatic accommodation reflex in the viewer's eye as it contracts and releases the ciliary muscles to bring the image into focus. But here one reaches the technical limits of the resolving powers of any optical instrument trained on an object made up from irretrievable data loss. If the observer drew the image closer to bring it into sharper view, they would find its form breaking down into tiny Ben-Day dots. Any attempt to recover the contents of its downgraded capture returns the viewer to this: a single, unyielding pixel—a visual signature of the informational absence withheld beyond a threshold of resolution.

The image on this book's cover enacts and opens the material, technological, and optical dynamics of resolution traced in this book. Resolution: the practice of transmuting an abstraction into form; the optical technique of coaxing something diffuse into clarity; the dissolution of a whole into its smallest elementary unit; the process of dispersion; the practice of making compatible. From these manifold definitions, *Microbial Resolution* unfolds the work that resolution performs in the U.S.-led war against the unfixed, unknowable, and immaterial futures of "emerging microbes." This book develops questions from the fields of visual culture and media studies to build upon a body of interdisciplinary scholarship on the global management of contemporary pandemic risk. *Microbial Resolution* is about thinking through resolution as its various dynamics are put to work in that U.S.-led effort, beginning in the late 1980s, that seeks to render emerging microbes into high definition and to place them at the center of a new global knowledge project. This undertaking—one of planetary consequence—stages the dynamics of emergence (as immaterial, mutant, unknowable, unfixed) in counterpoint to the mission of resolution (a drive to materialize, fix, clarify, analyze, and solve). I call this project "microbial resolution" to mark its drive to solve the problem of emergence, while retaining the etymological root of "resolution" that gestures to the interminable nature of this effort (*solver*: dissolve, break-apart).

The questions motivating *Microbial Resolution* began long before Covid-19 was in anyone's sightlines. The book was completed during the pandemic. Readers who lived through that time may be inclined to search for ways to understand it in the following pages, and may find some conceptual guideposts to help them sort through that experience. However, *Microbial Resolution* does not set out to offer an analysis of the Covid-19 pandemic. Nor does it take up the other pandemics and outbreaks that have appeared and receded since its inception (for example, the swine flu pandemic in 2009, or the avian flu outbreaks of 2013 and 2017). All of these events came with their own vortexes of 24/7 news cycles, affective noise, and biosecurity buildup, followed by a gradual and often indeterminate drawdown. Each has tested and refined the book's argument. But my aim in this book is to reveal the structure of the broader project of microbial resolution that underpins such events, and so it has been important to decipher the stable line of continuity that moves through, and is obscured by, the experiential turbulence of such moments.

To do this, I kept centered on resolution as an analytic and an optic as I developed the book. Keeping in mind that resolution can refer to the way an eye shifts planes of focus—allowing objects in the foreground to blur as the background comes into view—I situate the war against emerging microbes in intertwined historical and sociocultural contexts: the formation of a new scientific knowledge object when post-Newtonian theories were reframing world systems as complex, nonlinear, and fundamentally uncertain; the search for new rules and modes of warfare in a historical moment that, for many, signaled the possibility of an era of unprecedented peace; the restructuring of vision amid new speculative ways of seeing, envisioning, sensing, and relating to the blasted-open space/time of contemporary life; and the rise of a nearly theistic faith in information as the primary medium through which all global systems and objects would be ordered, understood, experimented with, and managed.

There are multiple historical accounts that unfold in the following pages that exceed any single moment. If I were to offer current and future readers a way to enter this book, I would invite them to reflect on the broader contexts that have shaped the project of microbial resolution, and the questions that they open up around the contradictions and aporias that structure the relation between vision and knowledge in the twenty-first century. If we keep our eyes trained on that unresolved space, there is, finally, much to see in that single pixel: what it brings to form, what it dematerializes and withholds, and what it leaves open.

Introduction

Microbial Resolution

Microbial Resolution maps out a crucial period from 1989 to 2016, when microbial life was reconceptualized and materially reworked. In 1989, a group of U.S. government scientists met to discuss some surprising findings:[1] new diseases were appearing in the world, while viruses that were thought long vanquished were resurfacing. Following the World Health Organization's (WHO) 1980 declaration that smallpox had been eradicated and the success of the global antimalaria campaign, the United States largely considered deadly infectious diseases a thing of the past. American public health was geared toward addressing chronic illnesses such as diabetes and heart disease. However, the AIDS epidemic and the appearance of multi-drug-resistant forms of tuberculosis proved this assumption incorrect. At the same time, the biological sciences revealed that microbes were far more complex than previously thought, and could no longer be understood as static, isolatable, and eradicable. With the Cold War nearing its end, U.S. scientists and security experts turned their attention to more amorphous and diffuse threats, including the risks posed by new patterns and formations of microbial life. It was in this context that they categorized microbes as "emerging" entities.

"Emergence" names a conceptual shift, not a pathogenic one. The emerging microbes concept characterizes microbes as dynamic entities, endlessly proliferating and perpetually mutating as they traverse global systems and ecologies. "Emerging microbes" have the capacity to appear anywhere at any time, and without warning. The concept invented by those U.S. scientists loads pandemic risk into a looming but constantly deferred future of unfixed and unyielding microbes. Although microbes had been understood as mutable since René Dubos initially described them in such terms in the 1950s, they underwent a profound resignification in the post–Cold War period.[2] While germ theory has long identified microbes as a "them" versus "us," the idea that microbes could reappear or emerge set humans and their technologies in a warlike contest against microbes.[3] U.S. scientists and biosecurity specialists understood this as a war of existential proportions, an "evolutionary competition" between "our wits" and "their genes" in which "the survival of the human species is not preordained."[4] "Emerging" here meant novel and threatening entities that strained the frames of established patterns of anticipation: new diseases hitherto unknown to humankind or ones that resurfaced after an apparent period of eradication. Yet as Paul Farmer pointedly reminds us, most emerging microbes are not "new," since many of them have long existed in other countries; they are considered emergent only when they begin to impact more visible or "more 'valuable'" regions and peoples.[5] For the U.S. security state, this new framework of emergent microbial life heralded a post–Cold War world of pervasive threats to health on a global scale as well as potential dangers to national and international security.

This book examines the temporal, scalar, and ontological reorientation of infectious disease management that attended this new framework. Once pathogens were characterized as "emerging," countering them was no longer a question of achieving national or global health, but of *securing* it.[6] This qualitatively different approach entailed a temporal reorientation as efforts were redirected from strategies of prevention to preparedness and preemption. This change in turn shifted the administration of health from national registers to global governance.[7] No longer were human populations its object; rather, the global security effort sought to encompass all living species and organic and manufactured systems, concerned as it was with the interdependent conditions of time, matter, life, and systems across the entire planet. This shift toward global governance was accompanied by the proliferation of a discourse on emerging microbes through such institutions as the National Institutes of Health (NIH), as well as in an array of journals and laboratories. How did a contentious concept developed in U.S. government conference rooms gain worldwide traction to become, in time, a "universal" scientific discourse? In 1992, the NIH published a key consensus study, *Emerging Infections: Microbial Threats to Health in the United States.*[8] This publication brought the emerging microbes concept to a broad audience for the first time. Alongside the sudden growth of journals, laboratories, and projects dedicated to emerging microbes, the U.S. Centers for Disease Control and Prevention (CDC), the NIH, and the U.S. Department of Defense reenergized institutional relations first formed during the Cold War.[9] The WHO, then in a period of institutional crisis, would renew its direction and reestablish its credibility in part by spearheading a global response to emergent microbial challenges.[10] The UN, in turn, called for refashioning the International Health Regulations to address emerging microbes. A U.S.-led project to counter emerging microbes globally thus gained transnational traction, eventually becoming a vast formation that spanned states, worked across international institutes, pervaded world ecologies, and took hold of global systems and resources. Its effects saturate the planet and shape its futures in ways that have yet to play out.

Resolution, a Transmuting Logic

The history of emerging microbes hinges on the problem of their materialization. How did those invisible and untouchable, omnipresent yet elusive potentialities become the material substrate for a planetary remaking? In the microbiological sense, a microbe is any microorganism, any minute form of life. The concept of emergence extends the threat of microbes through an unending chain of mutations. But the discourse of emergence did not originate in molecular biology, nor was it limited to its study. The rise of the emerging microbes concept was deeply shaped by broader historical processes through which complex systems science came to prominence in the 1980s. The rise of complex systems science (or the science of chaos) marks a rupture with the predictable dynamics of Newtonian laws.[11] The ordered Newtonian world is ruled by linear dynamics of cause and effect, develops in states of equilibria, and remains homeostatic. In contrast, the science of chaos sees the world as governed by complex systems where dynamic interconnections perpetually give rise to self-organizing systems; discrete components come together in nonlinear and unpredictable ways, out of which emerge new and increasingly complex systems.[12] In place of the ordered, symmetrical, and continuous linearity of the Newtonian world, the signatures of chaos science are disorder, irregularity, discontinuity, and fractal growth. While its systems are illegibly complex, the science of chaos views them as nonetheless expressive of a hidden deep structure. A larger science of "emergence" developed in the natural and social sciences from these ideas that took the future—as a realm of always incipient unforeseeable change—as an object of knowledge production.[13] Studying emergence defied previous modes of inquiry and comprehension,[14] and experimentation, prediction, and projection became the primary modes of analysis. The principles and methods of emergence theory were installed across a range of fields, including microbiology.

The discourse of emerging microbes and the task of securing their endlessly uncertain futures captures the renewed mentality of U.S. national security in the post–Cold War era. Securing microbes meant disclosing their always emergent and never stable futures. This book homes in on a contradiction inherent in what U.S. health and security experts came to refer to as "the war against emerging microbes":[15] How can the unstable horizons of emergence be secured? By rendering *unseen, unknowable, and unforeseeable* possibilities into *visible, apprehensible,* and *calculable* forms. I deploy the concept of *resolution* in an analytic framework that probes this process as a materializing logic. I open up the term in its optical, temporal, juridical, technical, and epistemological senses. In doing so, I examine how resolution works in counterpoint to emergence by performing its transmuting functions, moving the world between a range of states and conditions. Resolution means the process of dissolving something massive into its most elementary unit in analysis, the practice of converting an abstraction into apprehensible form, a mechanism of dispersion, a feeling of determination, a solution to a problem, a solvent, the movement from dissonance to consonance, the technique of making commensurate, the expression of legislative will. I use the term *microbial resolution* to refer to two crucial, intertwined dimensions that make up the war against microbial emergence. First, the term gathers a diverse range of discursive processes and visualization practices in the U.S.-led war against emerging microbes. Here, resolution also works in its meanings as solving a problem, as bringing an ill-formed entity (such as a diffuse threat) into shape, and as a spirit of determination in a project formed out of U.S. post–Cold War national security commitments and priorities. Second, I use the term to characterize this project as dependent on the ability to transform "emergence" into a workable epistemological object through processes of visualization and mediation. In this sense, microbial resolution refers to the set of discursive processes and visualization practices through which emerging microbes are coaxed out of the realm of discourse and transfigured as material forms within the world.

Microbial Resolution, a Problematic of Vision and Mediation

In the humanities and social sciences, scientific knowledge production is treated as a matter of representation, of uncovering the concealed or visualizing the invisible. I contend that the task of securing microbial emergence needs to be understood not as a problem of representation or sight, but of resolution. But the process of emergence moves directly counter to resolution; it resists closure and exists in the absence of definition. Emerging microbes cannot be brought into focus through optical lenses that enlarge the material facts of entities but never extend past them. Microbial resolution traces the complex problematic of apprehending emergence: How do you see something that does not (yet) exist? How are the notional and unknowable futures of emerging microbes transformed into visible scientific objects? In addressing this question, microbial resolution pushes aside optical limits to cultivate new modes of seeing within the computational, the molecular, and the sensorial.

A visual methodology is crucial for understanding the microbial discourse that developed between 1989 and 2016. Because this discourse characterizes microbial threats as ever present (but always deferred), in order to fight them the United States would need to develop strategies beyond deterrence. Rather, as the authors of *Emerging Infections* state, the United States should work through global institutions to "take the lead in promoting the development and implementation of a comprehensive global disease surveillance system."[16] Such a system would allow the United States to "get ahead of the curve of emergence."[17] Here, the National Institutes of Medicine (IOM) recommends reviving and amplifying the biodefense infrastructures of the Cold War, and developing a broad range of new anticipatory techniques capable of making microbes knowable, operable, and governable objects in the present. This militarized project to foresee emergent microbial futures spurred the need to expand the capacities of visualization and mediation—materially sensorially, technologically, infrastructurally—so that emerging microbes might be anticipated and secured before they materialized as probabilities. But the process of microbial resolution tests and strains these capacities, setting loose a series of problematics at their peripheries.

This book examines how techniques and technologies of visualization and mediation—such as data visualizations of microbial ecologies, computer-modeled futures, GPS-tagged animals, gene sequencing, and aerosolar particle modeling—were developed and deployed around the U.S. project to render the futures of emerging microbes manifest in the present. Linking these techniques and technologies to larger material, sensorial, affective, and political ecologies, I examine the mutant epistemologies, ontologies, and environments that surface through strategies of microbial resolution. I ask: What precisely does it look like to draw always incipient microbial futures into being as operable territory? And if emerging microbes are ultimately unfathomable, what does it mean to coax them into knowable and graspable form? How did the project of microbial resolution harness media forms, imaging techniques, and material and virtual networks to manage the world in an effort to secure wealthy nations from pandemic threats? This specific orientation to the visual concerns the multiple questions of resolution, and so what follows is not primarily an analysis of visual and narrative representations of contagion in media. Compelling studies in that field have already been carried out by other scholars; Priscilla Wald's *Contagious: Cultures, Carriers and the Outbreak Narrative* and Kristen Ostherr's *Cinematic Prophylaxis: Globalization and Contagion in the Discourse of World Health* are two noteworthy examples.[18] These works insightfully dissect the raced, gendered, and classed representations of global contagion narratives. While I engage with many aspects of this crucial scholarship, I focus on the ways that emerging microbes establish the conundrums and paradoxes that structure their mediation. In what ways do their obdurate indeterminacy shape the means through which we can know them, and even what knowing them comes to mean? A focus on emerging microbes finds meaning in materiality and a universe of affects, senses, and experiences outside, alongside, and even despite the meanings and hermeneutics of representation. Without neglecting the latter, this book advances a materialist, archaeological, and ecological analysis of vision, visuality, and mediation. In doing so, it examines how the project of microbial resolution recasts the very senses of vision, perception, and observation.

"Resolution Determines Visibility"

"Resolution determines visibility." Hito Steyerl repeats this pronouncement throughout her video, *How Not to Be Seen: A Fucking Didactic .MOV File* (2013). The work focuses on determining who, what, and how something or someone becomes visible or invisible in our digital worlds, willingly or not. To evade the all-seeing gaze of the digital era, Steyerl offers viewers some humorous strategies for how to be unseen,[19] one of which is to "become equal to or smaller than one pixel." "Resolution determines visibility." Steyerl's assertion starkly summarizes the issues of power at the core of resolution. At stake are the means of controlling whatever becomes visible, and of mastering what may be erased from the domain of visibility: the removal of data trails from data banks or, more darkly, the "disappeared enemy of the state." Eyal Weitzman is similarly concerned with the politics of the pixel. Through his role in the Forensic Architecture collective, he grapples with objects, elements, and events that vanish because they are equal to or smaller than a single pixel. Yet these elusive images are often in urgent need of recovery as evidence in human rights cases. In one instance, Weitzman was called upon to prove the American bombing of civilian areas during Western drone strikes in Pakistan, against the assertions of the CIA. The Forensic Architecture collective investigated thirty-centimeter holes pierced in the roofs of buildings that had likely been struck by American firepower: the deadly signature of Lockheed Martin's Romeo Hellfire II, the CIA's munition of choice for urban targets. Yet such images failed to stand up as evidence precisely because the size of the punctures was less than the ground surface covered per pixel on public satellite imagery. The problem is one of incomplete resolution: because the hole matched or was smaller than the pixels that composed the image, it was impossible to bring into clear view.

In such cases, crucial pieces of evidence are blocked by the resolving limits of the technology through which they are recorded.[20] Frequently, the mechanical tolerances of these media are themselves controlled by human operators. While most public satellites are technically capable of achieving finer degrees of resolution, their capacities are deliberately limited for the "innocuous"

Figure I.1. Steyerl instructs viewers to "be equal to or smaller than one pixel" to escape the all-seeing gaze of digital capture. Screen capture from Hito Steyerl's video *How Not to Be Seen: A Fucking Didactic Educational.MOV File*, 2013.

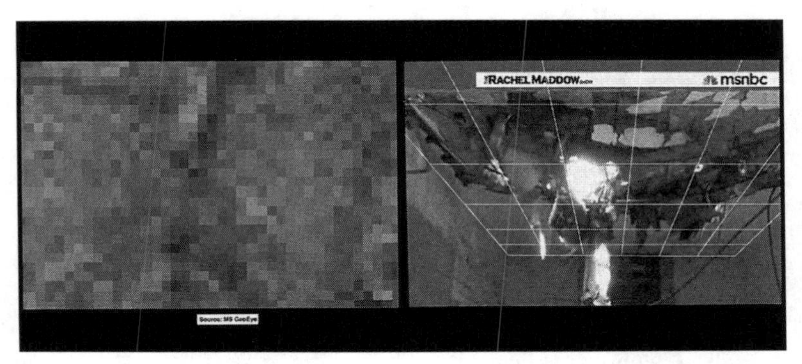

Figure I.2. Enlarged satellite image at the presumed location of a drone strike in Miransha, FATA, Pakistan (right). The Forensic Architecture collective was unable to verify whether a drone strike caused the hole in the roof because the hole is smaller than the size of a single pixel (left). Screen capture from a video report prepared by Forensic Architecture in collaboration with In Situ, 2014.

protection of privacy in the blurring of license plates, or, more insidiously, to conceal strategic military sites and violence from public view.[21] In this vein, Weitzman characterizes pixels as a kind of fishing net: objects are either large enough to be captured and rendered visible within its grid of intelligibility, or fall through the holes and vanish. Elements smaller than a single pixel drop below a "threshold of detectability."[22] By the same token, those who wield power over technical calibration also control what information can be captured or, conversely, what will be pushed into illegibility along the dissolving edges of the framing view.

I cite these examples because they remind us that there is a politics of resolution. Such instances underline that what is shown or withheld, what becomes visible or invisible, is determined by thresholds of resolution across the spectrum of all media.[23] Whatever cannot be captured by the technical affordances of resolution is drowned out in blurring smears and disintegrations. In scientific pursuits, media technologies convert a world of hidden things into visible definition: microscopes enable us to see across registers too small to behold with our eyes, while telescopes permit us to view detail at vast distances. In this sense, resolution is intimately connected to the epistemophilic drive that underpins the scientific gaze, and promises that what can be seen can also be known. This book examines what resolution crystallizes into knowledge and what it consigns to absence. Yet emerging microbes are unlike leaf patterns, organs, or other scientific objects insofar as they exist only in a state of ephemeral and perpetually mutating possibilities. For this reason, they leave the operation of resolution and its epistemophilic logic with a contradictory and open space. Here, emergence (dis)appears as a gap. Its condition leaves us with a problematic of visualization and mediation unaccounted for in earlier studies of resolution. What can be resolved when there is nothing to see?

If "resolution determines visibility," the project of microbial resolution is a quest to make emergence manifest, while constructing the instruments and processes capable of detecting it. That task presents a significantly different problematic from Weitzman's attempt to recover the absence of an existing hole. I examine how the process of microbial resolution retrains vision, visuality, and mediation around the speculative effort to see and secure microbes that do not yet exist. In that quest, the project of microbial resolution joins a range of scientific, political, and technological discourses, while reformatting vision and visuality through material, imaginative, affective, and legislative registers. As it does so, it forges curious links between vision, science, and security. In this way, microbial resolution trains our sightlines into the abyssal blankness, teaches us how to look at its emptiness, and instructs us on what to see in its openness.

Emerging Microbes and the Making of a New Global Knowledge Project

The U.S. project to secure emergence forms the starting point of a new global knowledge project. However, my argument is not that the post–Cold War American security state became somehow capable of injecting the world with its national interests and agendas, nor that the project unfolded with homogenous effects. As it became a global formation, the emerging microbes concept was necessarily entangled in the interests, frameworks, and agendas of many states, agents, laboratories, and institutions. For example, managing microbial emergence may be understood as the territory of microbiologists and virologists, but it relied as much on networks of ecologists, logistics specialists, small- and large-scale farmers, food systems engineers, data experts, and others; although rooted in the imperatives of a post–Cold War U.S. security state, mobilizing the project required discussion with international agencies and negotiation through their member states, and often faced contestation and debate. These global political and epistemological contexts dislocate, decentralize, and reshape the national project. So while the United States is a hegemonic state, the point of this book is not to illustrate that hegemony per se. Instead, *Microbial Resolution* approaches the making of a new global knowledge project as a problematic of historical epistemology. It attends to the epistemic systems through which "novel objects come into being and are shaped in the empirical sciences"—what Hans Jorg Rheinberger describes as the interactions of the historical, technical, and cultural dimensions of scientific experimentation.[24] This approach affords a vantage point distinct from a hermeneutically suspicious or a purely social constructivist stance that might hold, for instance, that scientists studying emerging microbes design experiments in ways that conform to (and either corroborate or disprove) existing theories. At the same time, it refutes an objectivist position of scientific knowledge, which asserts that truth is out there in the world as some kind of "empirical prey" to be caught by science.[25] It also prevents the misconception that scientists and experts working on the effort to manage emergence have done so in alignment with the particular ideological positions of the post–Cold War U.S. security state. Instead, I approach scientific knowledge from the understanding that its historical and cultural contexts form the "conditions of possibility" through which various experts work and experiment, and shape the fundamental concepts that "organize the knowledge of different historical periods."[26]

Microbial Resolution tracks how the notional figure of emerging microbes surfaced as an object of global scientific knowledge and public discourse precisely at the point when political and economic relations, technology, infrastructure, as well as the understanding of earth systems were being reconfigured in the new post–Cold War period. This was also when economic globalization and financialization restructured the world as a system that deepened financial inequalities while routing their consequences into a crisis-driven economy. At this time, theories of world systems recast the planet from a self-regulating, homeostatic entity to a space of constant disequilibrium and catastrophe, each exceeding scientific understanding and control. Technologies such as genetic sequencing, advanced computation, and informatics were also altering the conditions of observation while giving rise to new knowledge forms and techniques. These developments coincided with the rise of new entities like the "private-public partnership," the revamped multinational corporation, and the intensified global logistics network as they were reinscribing the world's infrastructure. Throughout this period, the United States went through a process of profound self-refashioning through institutions, technologies, and material and political landscapes, and rerouted its militarized energies, apparatuses, and economies to other platforms to engage the world and the planet through different means. This historical period forms the epistemological environment from which the scientific knowledge project around emerging microbes develops.

If the global war against emerging microbes is not a uniform and universal project, what is the value of an account that charts out its post–Cold War security origins? As the sole remaining superpower in the post-Soviet era, the United States was in a particularly favorable position of influence. The global knowledge project around emerging microbes was an integral part of this development. While the concurrent rise of multilateralism might have signaled an opportunity for international cooperation in global affairs, the United States instead exercised its newly uncontested power to shape global affairs, despite dissent. For example, during the 1980s and 1990s the United States applied financial pressure to the WHO to accomplish certain goals as the wealthiest member state to fund the agency at that time.[27] And while the WHO struggled to address the HIV/AIDS epidemic, the United States worked through their influence in bodies like UNAIDS, the World Bank, and the International Monetary Fund (IMF) in ways that repositioned the organization's standing. It also helped to launch philanthropic organizations with massive political, cultural, and economic clout in global health affairs, including the Bill and Melinda Gates Foundation, and later, U.S. government/corporation hybrids like George W. Bush's President's Emergency Plan for AIDS Relief. These entities further diminished the governing role of the WHO by economically hamstringing it, and then making it conform to the demands of the new neoliberal landscape.[28] In the 1960s, the WHO was the only actor on the world stage for global health. By the end of the Cold War, it was reckoning with "bilateral donors, multilateral donors, financial institutions like the World Bank, and mega-foundations like the Gates Foundation, and NGOs."[29] While the project of global health (to which the emerging microbes concept claims to give greater purchase) may seem grounded in a cosmopolitan concern for human welfare, it is deeply implicated in the imperatives of the post–Cold War security state.

Preemptive Biopreparedness, and Modes of Living with Risk

Emerging microbes were, from the outset, conceptualized as a new kind of national security risk. As such, the problematic of their resolution was defined as a matter to be addressed through militarized means. This securitized approach introduces a tension between the logics of emergence and the effort to resolve it. On one side, emerging microbes present risks whose scales of potential catastrophe are impossible to grasp and whose futures are unknowable. On the other, despite the proliferation of techniques and technologies to manage that risk, human efforts to understand, govern, and control it are proving insufficient. These two sides are intimately related. If "emergence" names the property of microbes as always proliferating and ever-mutating entities, then those microbes are not merely unknown, but *nonknowable.* The project of microbial resolution does not deal with the epistemological problem of making the unseen and unknown into visible knowledge. It reckons instead with the nonknowable. Moreover, its task is not to convert the nonknowable into something that can be cognitively grasped; rather, it operates to make the hazy frontiers and limitless dangers of microbial emergence into a productive and operable object. Instead of seeking to overcome the limits of knowledge, then, the project of microbial resolution enfolds nonknowledge into visual, political, and material relations in ways that thread the world through with its precarious charge.

This book draws from scholarship in critical security studies to analyze how the project of microbial resolution works through preparedness and preemptive logics to channel radical uncertainty and precarity as means of power and administration. The effort to "make risk work" appears in the interplay between these two logics. I track how "preemptive biopreparedness"[30]—the quest to build readiness for incalculable catastrophic pandemics by acting on their futures in the present—depends on resolving the nonthing of microbial possibility in ways that shape material, epistemological, and affective realms according to specific biosecurity configurations. The emerging microbes concept prompted experts to move away from strategies of disease prevention and toward anticipatory techniques of preparedness. Whereas *prevention* addresses known and existing dangers, *preparedness* addresses the unknown future. These modalities of future governance are also distinct from *precaution,* which seeks to control determinate future threats (particularly environmental) by acting on them before they become insurmountable in order to avert catastrophic yet uncertain futures.[31] Preemption, by contrast, acts on indeterminate futures rather than on a defined hazard in the present. Unlike prevention and precaution, which seek to deter bad futures, preemption is a generative technique, courting catastrophes rather than seeking to avert them. In this modality, dramatic interventions are made under conditions of immense uncertainty; because the catastrophic future threat is so nebulous, momentous action in the present becomes necessary. The point is not to stop the catastrophe (per the precautionary approach), but to work within speculative modalities to bring about the futures that are its very target.[32] Because these threats are by nature incalculable, their effects far reaching, and their manifestations unpredictable, preparedness seeks to manage their potential fallout through strategies of institutional readiness, capacity building, structural resilience, and systematic stockpiling. As critical security scholars such as Andrew Lakoff, Ben Anderson, and Lindsey Thomas point out, the measures and policies of preparedness and preemption immerse us in speculative scenarios of ruined futures.

These forms of future governance work in the anticipatory mode, effectively staging the world as always teetering on the verge of a disastrous event; they do so by fashioning global geographies, institutions, infrastructures, and everyday life in the image of the disasters they seek to address. Preparedness and preemption make us live and materialize ruined futures in the present through fictional modalities.[33] These same techniques were mobilized in the war against emerging microbes. The invention of the emerging microbes concept was accompanied by declarations from U.S. experts that the nation needed to become prepared for a disastrous future pandemic. As Andrew Lakoff argues, these calls did not surface out of a sense that the nation had once been prepared and then became less so. Rather, only when health was conceived as a matter of preparedness did the United States feel unprepared.[34] As a norm of readiness came to structure how experts conceived of pandemic management, the goalposts of public health shifted; no longer something to be obtained or improved, public health became a thing that needed to be secured. When U.S. experts conceptualized emerging microbes as a national security threat, they began preparing for and preempting a perpetual war in which the future is always the object to be secured (even after actual pandemics or outbreaks have already occurred, such as SARS in 2003, swine flu in 2009, avian flu in 2015, and Covid-19 in 2020).[35]

In coaxing out the mechanisms, structures, and logics of preemptive biopreparedness, I do not analyze case studies of specific events (whether Covid-19 or the rise of the Zika virus in 2014). While most scholarship on pandemics and contagion dwells on the high drama around biological disasters, I focus on the dreary, "everyone everywhere" effort to manage generic influenza and flu-like respiratory pandemics to gain a view of the ways that this militarized health project profoundly reorganizes everyday life. Experts and politicians say they value preparedness exercises and techniques not because they promise to generate readiness around a specific event per se, but because the capacities produced can be transposed to other "generic biothreat[s]," as well as to other kinds of national emergencies. As Health and Human Services Secretary Mike Leavitt stated in 2005, "even if we are spared from a flu pandemic, the [preparedness] work we do today will serve us all well in the event of any national emergency."[36] Rather than analyzing how specific events are or have been handled, I maintain this genericism to draw into view the mechanisms through which the project of microbial resolution renders "always-already ruined futures" as an ever-present backdrop and the permanent material and political conditions of contemporary existence.[37]

Despite—or because of—its efforts to secure an unknowable future, the project of microbial resolution is marked by heterogeneity and doubt. Rather than resolving this heterogeneity, I maintain its incoherence to explore the vexing terrain of a planet saturated with unyielding and uncontrollable risk. In place of generic notions of danger and threat, however, I deploy the concept of risk as intricately woven into the complex infrastructures of modernity, including global supply chains, food manufacturing and distribution systems, logistical planning structures, insurance estimates, global travel, and trade.[38] The framework of risk foregrounds complex questions of temporality, knowledge, and scale in relation to emerging microbes. I take many cues from Ulrich Beck's theory of risk, even while departing from it at critical points. In Beck's view, risk is a problem specific to accelerated modernity; the very processes that constitute our lives today are the same forces that infuse them with systemic vulnerabilities.[39] We teeter perpetually on the edge of catastrophe, and when those fateful moments occur, as they now and then must, they unfold across scales of incomprehensible temporal and spatial magnitude. Scholars such as François Ewald and Frank Knight, in turn, have theorized risk in relation to insurance and as subject to actuarial calculation.[40] Yet the material logics of emerging microbes slip through that precise measure. When, where, and how they might arise cannot be foreseen with certainty by any expert. For this reason, I draw primarily on Beck's conceptualization of risk as a realm of the unknowable—"incalculable," "unforeseeable," "immeasurable"[41]—while acknowledging and analyzing the considerable efforts to apprehend the menace of emerging microbes. But whereas Beck's theory of "risk society" designates the rise of societies built around averting risk, I analyze the ways that the project of microbial resolution manages risk in order to render its uncertainties profoundly operable in the present.

By 2015 health agencies worldwide were emphatically recommending that war-like metaphors be dropped in framing human dealings with emerging microbes. It was generally agreed that such militarized metaphors encouraged the overuse of antibiotics and antimicrobials as counterproliferation tools which, in turn, led to the rise of drug and treatment resistant "superbugs." The rhetoric of war would be replaced with more symbiotic terms in an effort to shift cultural attitudes and to cultivate synergistic relations with microbes. Despite this shift, the frameworks of preemption and preparedness endure as powerful frameworks for human interactions with microbes and they continue to perform cultural and political work. While scholars have identified different, often competing, regimes of global health—humanitarian, militarized, biomedical, or social, for instance—the frameworks of preemption and preparedness are now so ubiquitous that it is increasingly difficult to think global health outside these models. Indeed, during the Covid-19 pandemic beginning in 2019, many people and organizations across political and ideological spectrums were pleading for more resources for preemption and preparedness measures to manage the pandemic and to prevent future ones. These frameworks have become so deeply entrenched in an overall understanding of microbial management that various regimes of global health, rather than contesting one another, have become more like different dialects of a model built upon preemption and preparedness (as chapters 4 and 5 will detail).

Emerging Materiality and Microbial Perspectives

What is the material status of an emerging microbe? Emerging microbes encompass their own contradictions, and so their status is necessarily ambiguous, oscillating between material and virtual, present and absent. Their material existence in the here and now points to unknown, unmanageable, and immaterial futures. Taking cues from work in elemental media, I pay attention to the ways that the material ambiguity of microbes structures vision and mediation and reformats our understanding of them. My treatment of media technologies grows out of the efforts of scholars across media and science and technology studies to restore materiality to technoscientific fantasies of information and communications as separable and divisible from the material forms in which they are embedded and embodied.[42] Premised on a Western liberal humanist split of mind/body and matter/information, the myth of virtuality aligns with the celebration of an ideal form and of universal man. The same valuations that view mind/information as the apex of existence and regard materiality as something to be shed, of course, also enable the multiple violences, erasures, and oppressions that result when real engagement with bodies and matter is elided. If "information lost its body" in "the flesh-eating 90s," as Katherine Hayles and Arthur Kroker suggest, a materialist perspective refurnishes historical and cultural contexts needed to bring its bodies and other matter back.[43] In the case of emerging microbes, a materialist analysis helps to reconnect their immaterial registers to the ecologies and bodies from which they are extracted, as well as the complex histories and processes through which they are brought into being. Despite the fact that microbes are actual material entities, the project of microbial resolution puts them to work as informatic and immaterial objects. Analyzing emerging microbes elementally means taking on the immaterial terms in which they are made to operate, even as I question the physically disembedding functions of this fantasy. How do the immaterial informatic dimensions of emerging microbes become "virtual spaces" through which otherwise intangible futures are thought to be accessed, and how are emerging microbes materialized as objects to be worked upon?

Along similar lines, I approach emerging microbes as media forms themselves. Microbes contain content; they pick up material across temporal and spatial scales, and record these accumulations in ways that remake their genetic constitution. In this sense, they are storage devices and instruments of material inscription. Microbes communicate substances between one another and to other creatures. They are transmissive devices. Just as with Marshall McLuhan's lightbulb (which changes a room by its very presence) or Rahul Mukherjee's infrastructures (which infuse our environments and bodies with radioactive danger), microbes radiate energies in ways that transform the spaces we inhabit and the ways we live there.[44] As Bishnupriya Ghosh shows, microbes shade things in and out of visibility by degrees of saturation.[45] Coexistence with microbes—on and within bodies, lungs, fingers, and surfaces—highlights the sense that the world is composed of various interfaces and content. This broad definition follows a different line of questioning than Eugene Thacker's theorization of biomedia, in which biotechnology reformats biology as computation—and vice versa—and draws theories of life together with concepts of engineering and design.[46] While taking this theory into account, *Microbial Resolution* treats emerging microbes as key elements mediating the realms of politics and culture in a nexus of social, racial, and regional relations.

This book also builds upon work in the environmental humanities that investigates anthropogenic change. Much of this scholarship draws out objects and processes too vast to see—processes of geoengineering, fossil fuels, capitalism, and war, for example—in order to reveal their implications in a world marked indelibly by human activity.[47] Yet we also need accounts of the minimal to explore the magnitude of planetary change.[48] This book sticks with the small. The question of resolution leads to the granular, the microbial, and the molecular. In examining how the war against emerging microbes works through minimal registers (for example, molecules or data "bits") in ways that reconfigure relations, ecologies, and processes, I recover some of the political erasures that arise when narratives of the Anthropocene stress massive processes. In examining how a project working on microbial registers took hold in ways that shaped global systems and infrastructures, I hold together the medical and environmental humanities, two areas often examined separately.

A focus on this minimal register also considers the forms of power hidden in the seemingly apolitical, standardized, and mechanical registers of technique and technology. Richard Dyer, for example, analyzes the technical limitations and affordances of color film whose technical baselines are administered to "light for whiteness"; lighter skin tones appear, while making black skin disintegrate into undefined "blobs" of darkness.[49] Across visual, media, race, and environmental studies, the power of resolution is important in representations of othering, and racial and colonial legacies. Their continuing afterlife has already been identified and richly theorized.[50] But it is essential to shift this concept beyond a history that treats it in largely representational terms. In making linkages between histories of information and computation, post–Cold War security paradigms, and the strange materiality of emerging microbes, I speak to the ways in which colonialism is recouched in other landscapes of resolution in the fight against microbial emergence. In order to open up the politics of resolution more broadly, then, I focus on the ways in which such relations have been transposed onto new terrains of information, precisely as other types of scientific representations—genetic code, data, quantification, and calculation—come together to form a new empiricism of power, control, and administration. An analysis of the minim enables a reading that locates inequality in the supposedly conflict-free realm of information, the solvent processes of quantification, the cosmopolitanism of a global health project, or the nonspecific geographies of microbial emergence. The colonial lines of disease and blame become diffracted through new particulate politics of air, data, and atmospheres, even as they spread across surfaces and become embedded in the earth's systems.

Chapters

This book is organized in two parts. "Part I: Aesthetics of Uncertainty" opens up the visual and material dilemmas at the core of the project of microbial resolution. These dilemmas are neither representational nor narrative in nature, but concern instead the optical and physical problem of coaxing the immaterial futures of emergent microbes into being as a substance sensed, felt, and lived through. They also involve how the contexts of post–Cold War politics with their historical, technological, and scientific dimensions reformat such aesthetic engagements.

Chapter 1, "Pathogenic Nation Making," documents the refashioning of twenty-first-century American nationhood and the discourse of U.S. preemptive biopreparedness in the face of microbial threats. I examine the media ecologies of *Deadly Migration*, a 2007

IBM documentary, as a case study of visual and material networks that stage the threat of emerging microbes as always potentially everywhere. In doing so, these ecologies produce "atmospheres of catastrophe": a world of warnings and reminders that transform lived, quotidian experiences such that Americans literally learn how to live *with* risk as a constant backdrop. The project to secure microbial emergence produces its affective and sensorial opposite: an ordinary, pervasive, and mundane insecurity in quotidian processes, objects, persons, and spaces. I use the term *pathogenic nation making* to characterize the refashioning of national solidarity based on mutual mistrust and the unbinding of social bonds.

Chapter 2, "Materializing Emerging Microbes," considers three of the defining narratives of the 1990s: the end of the Cold War, the rise of the genetic revolution, and the appearance of a new paradigm of global health. While these narratives have previously been considered three separate histories, I read them together to ask how it matters that the emerging microbes concept appeared at their intersection. In this context, the chapter examines how the immaterial and nonexistent entity of emerging microbes materialized as an object of scientific knowledge in the 1990s. I track its transformation from abstract discourse to matter in the world, from an immaterial threat to new post–Cold War national enemy, showing how a new war against a looming threat became a global campaign articulated through ecologies of planetary proportions.

"Part II: The Calculative Imaginary" focuses on the proliferation of and cultural investment in numeracy, quantification, code, data, and technological rationalities—what I call the *calculative imaginary*—that surfaces in efforts to bring emergent microbial futures into view. I theorize the calculative imaginary as a mode of vision that begins from the impossible optical imperative of seeing something that does not yet exist, and that conditions and underwrites the infrastructural and logistical reformatting of the planet.

In chapter 3, "Flightlines and Sightlines," I ask how *nonknowledge* and *uncertainty* surface as crucial elements in twenty-first-century epistemologies. I examine the use of migratory birds as data visualization technologies to foresee and forestall possible avian flu pandemics. I call this animal/technical assemblage *animal sentinel media* and examine the data visualized as risk maps that are produced from their bodies. Drawing on theories of risk, uncertainty, vision, and mediation to analyze the discourses through which these data visualizations are endowed with evidentiary force, I track a shift in the making of scientific evidence in twenty-first-century cultures of risk. This shift loosens the customary bind between sight and knowledge, establishing instead a relation between sight and foresight. What does it mean to make emergent events objects of knowledge through these risk maps? While such knowledge can only be replete with error, noise, and incompletion, these failures are central qualities in twenty-first-century knowledge systems. They give rise, in turn, to new modes of perception, techniques of observation, and foundations of biopolitical and planetary governance.

Chapter 4, "Fluid Economies of Biosecurity," considers how the view of microbial emergence as bioinformational plentitude was taken up more broadly to fashion an illimitable economy. I begin with the Global Viral Forecasting Initiative, a "virus-hunting" laboratory founded by scientist-entrepreneur Nathan Wolfe, to analyze an American technocratic fantasy of mounting a data-driven global pandemic prediction system. I examine how the United States deployed what I term the *language of harmonization* to promote the idea that all nations, regardless of economic ability, should equalize their scientific and technological capacities to meet new standards of global health. Information becomes the central medium in this new global project and is treated as if solely capable of ushering the world out of a range of crises. A global pandemic prediction system based on information demands that an uneven world be transformed into a smooth communicative platform, that life systems be reformatted as informational resources, and that data circulate perpetually in the virtual and material networks of global exchange and capital. I use the term *fluid economies of biosecurity* to describe these processes and to name the politics of informational movement and circulation called up in this paradigm of biosecurity.

The language of harmonization, however, harbors a vast range of contradictions. Chapter 5, "Managing the Microbial Frontier," brings these conflicts to the fore by reframing the bioeconomy as a problematic of logistics. I revisit the object and topic of the previous chapter, the Global Viral Forecasting Initiative and the U.S. effort to raise the world's capacity for global health, this time reading the regulatory landscape of the bioeconomy as a global logistical infrastructure that keeps bioinformation moving in a careful choreography of fix and flow. I analyze how this logistical system establishes a set of irresolvable contradictions at the core of the global health project; it counterpoints the imperatives of capitalism to those of humanitarianism, militarization to global health, and the economic to the biological/biospheric, rendering their divergent objectives coterminous. I examine how these contradictions are operationalized to craft new forms of synergistic tension between emerging microbes and emerging markets. The novelty of emerging viruses—their capacity to appear anywhere, any time, and in infinite genetic combinations—promises boundless economic production rooted in speculation on the perpetual potential of infected futures.

After all of the resolving work analyzed in these pages, *Microbial Resolution* shows that *irresolution* lies at the heart of the war against microbial emergence. If the project of microbial resolution volleys us across vast temporal and spatial scales, affective registers, dynamics of risk and security, near futures and their always-deferred catastrophic endings, it finally leaves us somewhere in between, suspended and irresolute.

Part I
Aesthetics of Uncertainty

1 Pathogenic Nation Making

A traveler flies home to the United States from a work trip abroad. Like many passengers who opted to view in-flight movies on select American Airlines routes between 2008 and 2010, this traveler sees a seven-minute documentary produced by IBM.[1] *Deadly Migration: Outsmarting the Avian Flu Virus* stages an account of why Americans should fear a potential avian flu pandemic. The video oscillates between projections of that catastrophe and possible methods to avert it. Struggling to open a complimentary bag of pretzels, the traveler watches distractedly while images of mass graves, panicked crowds, and postapocalyptic landscapes evoke widespread devastation and human-species extinction. Expert figures, surrounded by sleek technologies that are constantly on, testify that the United States must intervene here and now in the hypothetical futures of avian flu.

Preemptive biopreparedness seeks to render a nation capable of acting on unknowable but possibly catastrophic pandemic futures by positioning its individuals, infrastructures, and institutions in a state of perpetual readiness. If this preemptive project is a problem of governing unknowable viral futures, what do cultural objects such as *Deadly Migration* reveal about remaking the nation under the shadow of its pre-ruined destinies? A new mode of nationhood was being fashioned within the cultural and political project of preemptive biopreparedness. I call this process "pathogenic nation making"; the term designates the forging of a sense of national solidarity through pandemic fear and avoidance. Although other nation-states engage disease management with their own securitized logics, pathogenic nation making names a process peculiar to the U.S. national project. As Joseph Masco shows, people living under the post-9/11 counterterror state are "emotionally managed" across a broad spectrum of terror.[2] This administration of primal feelings, he argues, draws rhetorical, material, and affective resources from the Cold War effort to shape all aspects of American life under the prospect of collective ruination and in the shadow of the nuclear bomb.[3] The period covered by this book (1989–2016) is rooted in the same historical framework and the imperatives of the 9/11 counterterror state that Masco calls its "ideological fulfillment."[4] Pathogenic nation making extends the same logic, defining and redefining nationhood through the collective contemplation of possible microbial futures and their catastrophic endings.

Numerous scholars have established that nationhood is defined and redefined through discourses of disease preparedness. Priscilla Wald suggests that contemporary pandemics highlight the dense and interconnected worlds of accelerated globalization. They make us aware of the sprawling networks that link our own bodies to pathogens from "distant places," those far-flung and racialized and classed regions so often framed as the origins of infection.[5] The imagined nation is strengthened by securing its porous borders against incursion from ethnic and racial otherness, and protecting the vulnerable collective within these lines of demarcation. The prospect of contagion invokes the precariousness of the national community, while instilling in individual subjects the desire to fortify the nation to assure their mutual continued protection.[6] Wald and others shed light on how nationhood is articulated through the management of national borders and the substance of the body politic, and show that visual and narrative representations shape concepts of nationhood through the discourse of global contagion. Still, how do cultural objects like *Deadly Migration* become effective, not only in defining nationhood against representations of difference, but also by materially retexturing the affective fabric of daily life, a critical register of experience for the project of nationhood?

This process cannot be understood through an analysis of the representational world itself. I shift from the role of representation to consider the function that mediation and media ecologies play in pathogenic nation making. A media-ecological approach emphasizes the spaces, habits, situations, and materialities of mediation, yet does not neglect representation. On the contrary, it situates images and narratives in and through the larger objects, networks, and practices of which they are a part and in which they operate, making visible the otherwise illegible relations between them. Such an analysis opens up an understanding of how an entire web of objects, processes, scales, and discourses—the symbiotic relations between, for example, representations in public health warnings, hand-sanitizing pumps, and public health policies—constitutes a network of effects and affects central to contemporary national governance. By drawing attention to the ways in which representations operate through a larger network of objects, practices, and discourses, other registers of experience emerge—registers lost in analytic modes whose account of visual experience rests on imagery and depiction. Representation under pathogenic nation making must account for both the remarkable and the unremarkable manners of everyday fluctuations and mediations. Sometimes we pay attention to the fascinations and diversions of the world's representational spaces. But sometimes we do so in keeping with or against authorial intentions. Or, at other times we do not pay attention, or we do but later forget, or we drift along distracted and barely notice what we may be taking in. What would we find if we extended our understanding of the representational world to account for the everyday realities that shape our reception of cultural texts such as *Deadly Migrations,* realities like habit, forgetting, and inattention? What would the process of pathogenic nation making look like then?

This chapter documents how media and media ecologies played a crucial role in mobilizing national collective affect around the project of preemptive biopreparedness. Using *Deadly Migration* as a case study, I examine how affects of uncertainty move through visual *and* material media as they network in and around objects, spaces, processes, and bodies. In tracking the media ecologies of pandemic preemption and preparedness, I highlight the fundamental governmental forms that surface during this period. In this context, microbial resolution aids individuals to sense a world saturated with always possible pandemic risk, such that everyday people (and their everyday feelings) also become infrastructural to the pursuit of a nationalized security project. In order to make a nation on alert, media ecologies discipline the imagination so that microbial risk can be preemptively imagined as always possibly erupting from proximities and encounters with the objects, surfaces, people, and processes that make up and suffuse the spaces and rhythms of everyday life. "Resolution" in this sense can also refer to an affective state of conviction. In tracking the role of media ecologies in the project of microbial resolution, I examine how the opposite affect—a pervasive sense of insecurity—becomes the basis of a new form of collectivity. As I will demonstrate across the following chapters, rooting U.S. nationhood in a deep and ordinary uncertainty lays the foundation of a global strategy to manage microbes throughout the biosphere.

Pathogenic nation making occurs through the governance of the senses. Unlike anything as hard and categorical as "the law" or "policy," this form of administration operates and takes shape as a sensation of "shimmers," of intensities moving in, through, and across objects, processes, places, and bodies.[7] It impinges on the senses and accumulates as a vague force on the skin that does not yet crystallize as knowledge, thought, or action. In addition to operating as a representation, *Deadly Migration* is also a *material mediation* shown on a touchscreen in an airplane cabin traversing international borders. By understanding how *Deadly Migration* works in these two capacities, I situate its material mediations as part of larger networks of effects and affects within a broader media ecology. I demonstrate how its call to preempt unknowable viral futures implicates and works with apparently unrelated actants, effects, and affects. These media ecologies render fear and speculation communicable across publics by creating "atmospheres of catastrophe": a world of messages, warnings, reminders, and instructions that organizes collective affects around the vagaries of future disaster. Through these ecologies, senses of U.S. nationhood are steadily made and remade under the prospect of disastrously mutating microbial futures. Finally, I consider how anxious collective affects (which might also manifest as indifference, nonbelief, confusion, or irritation) become infrastructural to the project of preemptive biopreparedness, examining such emotive noise and confusion to reflect on the content and affective quality of the kind of nationhood that surfaces through these processes. As uncertainty inflects breath, infects touch, and troubles proximity, pathogenic nation making brings about forms of nationhood rooted in a sense of social fracture.

Deadly Migration and the Discourse of Preemptive Biopreparedness

Deadly Migration weaves together two stories. The first is the tale of a deadly pandemic that originates in China and hovers on an ever-nearing future horizon. The second communicates the source of the threat: viruses endlessly emerging into ever-more dangerous strains. From a virus to a swan, from the documentary's fictional character Ming Xio to his cousin, *Deadly Migration* outlines the quiet and all-too-easy transmission of biological material from one organism to another. The documentary literally connects the dots between animal and human bodies, thereby collapsing them as it lays out the hypothetical scenarios that might lead to a global avian flu pandemic.[8] By graphically rhyming and multiplying shots of the virus with birds and people, *Deadly Migration* establishes their equivalency as coconstitutive elements of viral emergence and vectors of contagion (Figure 1.1).

The seamless conduits of viral transmission are highlighted in red (Figure 1.2). Commencing with a graphic of a scarlet swan over a map of China, a detail that isolates the presumed origins of the infection, the virus eventually latches onto Ming Xio's engineer cousin, dressed in red at an airport gate. It then takes flight. In the next shot, a small red dot on the pristine white surface of an illustrated airplane indicates an infected passenger on board. In the following sequence, identical dots seep through the rows of seats to show how the virus spreads to all other passengers, until a ruddy halo encircles the plane. As that mechanical bird of contamination deposits its infected passengers on the map of the northeast coast of the United States, a voice-over announces: "Now the humans are the problem." A red circle radiates outward from New York City. At the same time, amber lines sprout from the metropolis in every direction as passengers travel on. Soon the

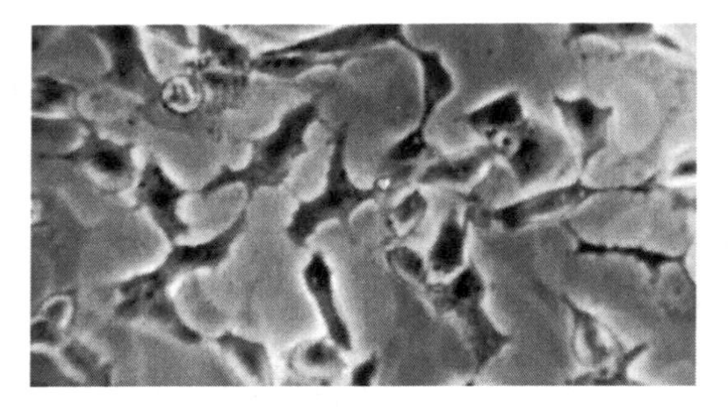

Figure 1.1. *Deadly Migration* graphically matches images of viruses, birds in flight, and crowds of Asians. Screen captures, *Deadly Migration: Outsmarting the Avian Flu Pandemic*, IBM Films, 2008.

world is awash in red. A soundtrack of eerie violins emphasizes the horror that borders and boundaries—species, race, life, and nation-state—no longer matter.

The visual commingling of crowds of Asians, swans in migratory flight, and mutating viruses implies that the bodies of faraway peoples, the biologies of distant animals, and the spaces of viral mutation are one and the same. *Deadly Migration* of course builds upon the well-worn narrative of the "yellow peril" to construct its narrative of global contagion, in which Asians infiltrating the globe

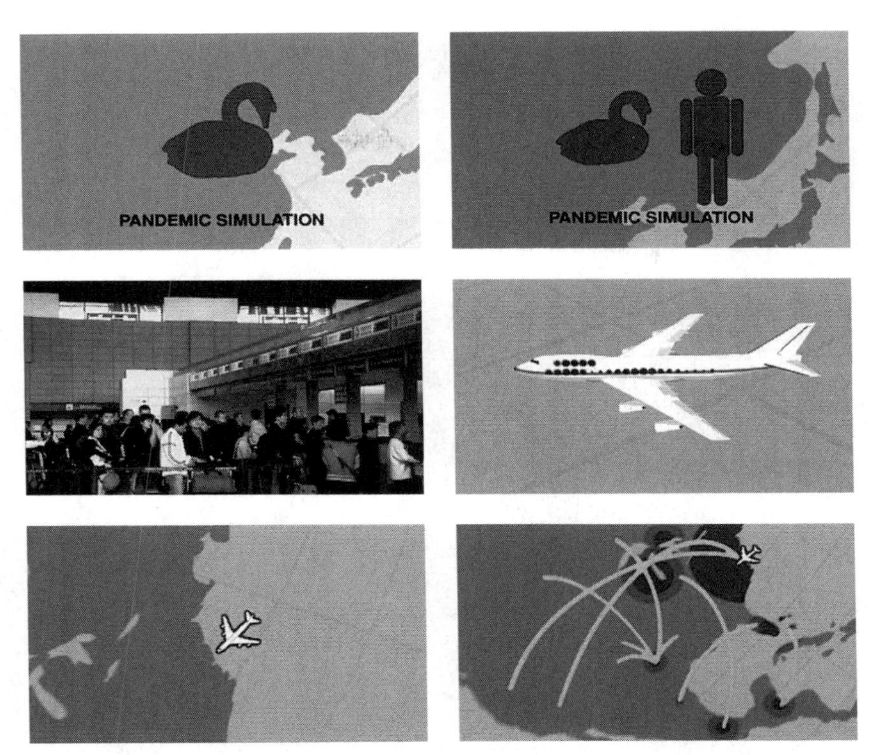

Figure 1.2. *Deadly Migration*'s pandemic simulation highlights possible pathways through which avian flu could move from a swan to a global pandemic. Screen captures, *Deadly Migration: Outsmarting the Avian Flu Pandemic,* IBM Films, 2008.

pose a direct threat to the United States. While many scholars insightfully address the racialization of contagion narratives, this documentary's story of racial overflow is part of the broader work of preparedness discourse in nationalizing global risk. As Lindsey Thomas argues, preparedness discourse, despite its presumed universality, addresses itself to the defense of "the homeland" and works to protect its presumed whiteness.[9] As the domestic component of U.S. counterdefense, it seeks to regulate the nation-space and render it ready by managing global risks. The documentary sets this worldwide scene by underscoring the global networks of infrastructures, economies, spaces, objects, and bodies through which microbes move to reach the United States. In charting these connections (and in rhetorically collapsing Asians, swans, and mutating viruses into one), *Deadly Migration* reveals the web of relations through which lives, objects, temporalities, and spaces are unavoidably interwoven and, ultimately, through which microbes penetrate bodies in the United States. We usually live largely unaware of the networks that constitute our daily worlds, but in tracking the courses through which pathogens may one day spread, this documentary shows that the "stuff" of everyday life is filled with "absent presences."[10] These "living but absent" and "dead but not gone" forces suffuse the microbe and charge it with haunting potentialities.[11] Illustrating RAND's claim in *The Global Threat of New and Re-emerging Diseases* that emerging microbes will ruin the American dream and the individual pursuit of "life, liberty and happiness,"[12] the documentary opens by presenting archival images of the United States during the 1918 influenza pandemic. A somber funeral dirge accompanies footage of the Boston Liberty Loan Parade. The video then cuts to photographs of victims of the pandemic. Later, over footage of U.S. soldiers fighting in the First World War interspersed with yet more pictures of sick and dying people, David Jasinski of the IBM Watson Labs informs us that the 1918 flu "killed more soldiers from the American army than all the fighting in World War I." These aspects, together with the documentary's ultimate concern that American cities will be threatened by the future of global contagion ("It spreads to New York, Boston, L.A., Miami"), reveal the United States to be the true subject of this story of global risk.

Because the world is so thoroughly interconnected, the documentary suggests that catastrophic pandemics inevitably await. But by virtue of these same complex connections, it is impossible to determine where, when, and how they might appear. In the face of the unknown, "the time to act," one of the documentary's experts tells us, "is now." And if we do not? A hypothetical death toll flashes on screen. As the numbers rise to the beating sounds of a racing heart, viewers are alerted to its mortal consequences. The magnitude of that future biological catastrophe is underlined by a shot of a peaceful mountainscape that shudders, in the blink of an eye, under an apocalyptic flash (Figure. 1.3). Human biological security itself is at stake. Through the sound of a chicken clucking

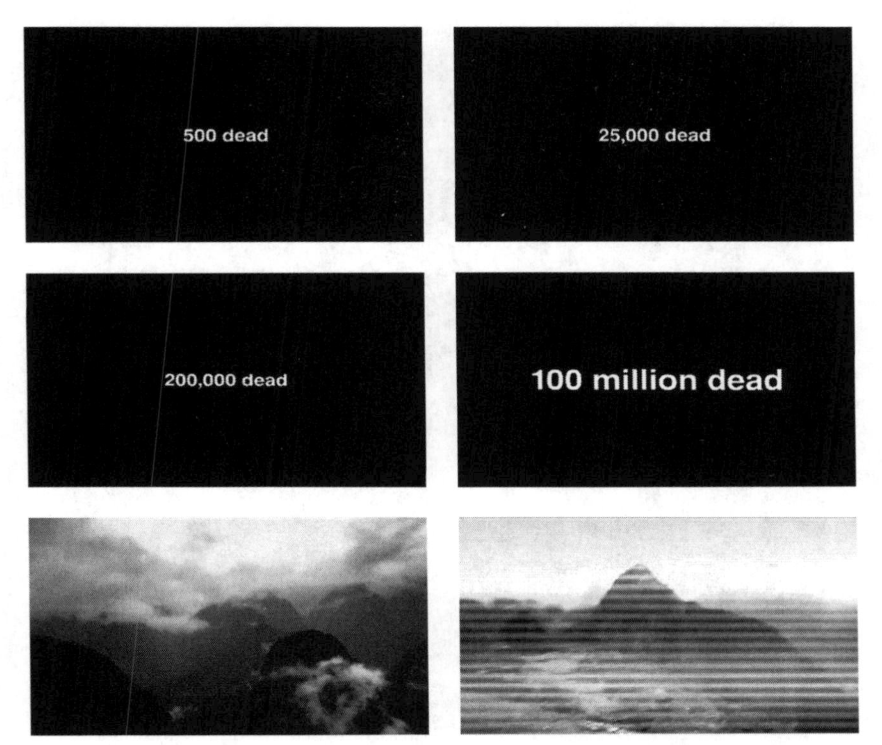

Figure 1.3. The numbers of a rising hypothetical death toll appear in quick succession, followed by a peaceful mountain scene that whites out under an apocalyptic flash. Screen captures, *Deadly Migration*'s "doomsday scenario," *Deadly Migration: Outsmarting the Avian Flu Pandemic,* IBM Films, 2008.

over the explosion of light, the sequence implies that the future may very well be one in which avian (non-American) viruses will emerge victorious. The flash suggests that the single, quick twist of viral transmission will occur without warning, at the speed of light, undetectably and irreversibly.

In order to preempt this catastrophe and get "ahead of the curve of emergence,"[13] U.S. experts intone, we need to develop a set of "anticipatory technologies" capable of seeing, predicting, and managing the futures of viruses and bacteria.[14] As Bill Clinton stated in a 2003 speech, the United States needs to cultivate anticipatory techniques and technologies that can detect emerging

Figure 1.4. An IBM expert declares that we must detect the "how" and "when" of hypothetical future pandemics. Screen captures, *Deadly Migration: Outsmarting the Avian Flu Pandemic,* IBM Films, 2008.

microbes before they become manifest risks because "you can't zap [them] with missiles, and night-vision goggles won't protect us against them."[15] Nicholas Tsinoremas, one of the experts featured in *Deadly Migration,* appears to agree. As he sits bathed in the cool blue glow of his computer screen, processors humming in harmony with his voice, he implores us to develop the technologies that enable imagining "HOW" and "WHEN" the coming catastrophe will unfold (Figure 1.4). In its urgent call to address hypothetical pandemic futures, *Deadly Migration* articulates a core paradox of preemptive biopreparedness: it regards the future pandemic catastrophe as unavoidable—"not a matter of if, but of when." Precisely because its futures do not yet exist, and are therefore beyond the reach of the usual modes of scientific apprehension, it is imperative for the United States to intervene in and secure these unknowable futures in the present. *Deadly Migration* participates in this broader national security project to secure the unknown; its mutation narrative frames the future as a never-ending frontier of risk that demands intervention today. One of the documentary's experts leaves us to imagine the dark consequences of inaction: "If we do nothing about bird flu and we knew this was a *possibility* . . ." Here, the documentary engages in the "politics of possibility," "the governance of low-probability/high consequence" events and their unknowable futures.[16] In contrast to the logic of probability (a form of calculation that seeks to identify the most likely futures), possibilistic reasoning envisions the future as a latent array of proliferating outcomes, none of which can be precisely predicted. The project of preemptive biopreparedness is thus a self-regenerating operation in *unyielding* combat with the always-possible, ever-present future pandemic.

How to make the unknown future appear? In the project of preemptive biopreparedness, these manifold futures can only be brought into being as manageable objects by thinking in fictional modalities. If the goal of preemptive biopreparedness is to ready its subjects for incalculable, unforeseeable, and unknowable pandemic disasters, the project must "inspire preparedness" in everyday Americans, in the words of Health Secretary Leavitt.[17] This means enabling subjects to experience the possibilistic horizons of the future by submerging them within its fictive modalities and prompting them "to think the unthinkable." *Deadly Migration* develops the shape of this blank future using various narrative strategies: it constructs archetypes like villains and heroes, haunts the future with the ghosts of history, and invokes conventions of horror and suspense. Most significantly, *Deadly Migration* shapes this unknown future through what Thomas calls the "training and preparing" genre. In this mode, national security projects build fictional futures to think through them and develop expertise—in Thomas's phrase, the "empiricist epistemology of fiction."

Deadly Migration trains its viewers to think the unthinkable through a fictional scenario-building structure. The "doomsday scenario" in which Ming Xio infects the entire world continues later in the documentary, and is clearly aligned with the speculative exercise outlined by the U.S. Centers for Strategic International

Studies: to imagine situations that are not the "most likely" but rather the "worst possible."[18] This kind of scenario-thinking, as Thomas demonstrates, aims not only to prod its subjects to imagine the future, but to discipline their imaginations so that they experience the present *as if already living* in the made-up futures of crisis. Accordingly, the use of what-if scenarios in *Deadly Migration* renders possible futures into ones in which viewers can place themselves. As the documentary cuts rapidly from images of swans to crowded urban centers to footage of global travel and trade infrastructures, yet another expert delivers a barrage of questions: "What if this? What if that?" the expert asks in sketching out various scenarios: "What if I manage to quarantine all the people . . . What if one escapes? All it would take is one. . . . This could end very badly." These what-if scenarios materialize an array of tangible possibilities out of the otherwise blank future.

A frenzy of calculation accompanies these what-if scenarios to lend a sense of scientific credibility to *Deadly Migration*'s speculations: images of projected numbers, computer-modeled bird migration pathways, simulated air and ground traffic flows, and rainbow gene proteins proliferate (Figure 1.5). The documentary then returns to a map, plotting projections onto a visual and traversable space and time; like military tabletop exercises that function in part to materialize the contents of a militarized imagination, hypothetical futures are visually anchored to this cartography: the dead are tallied, potential nodes of transmission are rendered real, and projected movements of unfriendly viral blooms are charted.[19] Because the "how" and "when" of emergent futures cannot be known, their manifold possibilities must be given shape by enumerating generic and "formulaic combinations of description and planning considerations" around those fictional futures: lists of places and things to watch for, estimated fatalities, calculations of infrastructural damage, approximate costs, projections of resource demands and shortfalls, and so forth, laid out in sequence and narrative form. The use of fiction to build scientific and social knowledge around emerging microbes in *Deadly Migration* collapses the difference between empiricism and speculation.[20]

Despite recognition that these modelled futures will inevitably be inaccurate, *Deadly Migration*'s what-if scenarios are distinct

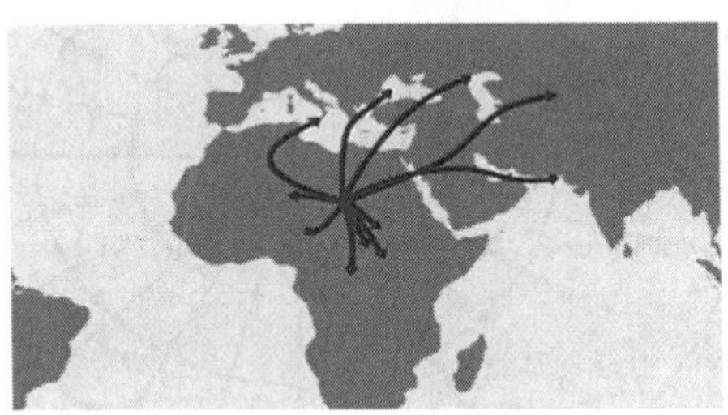

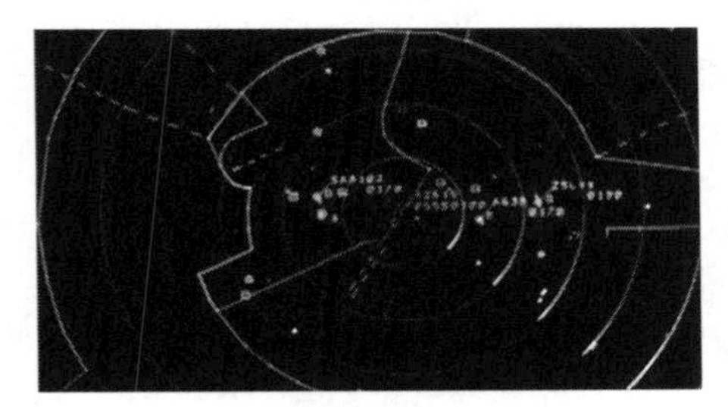

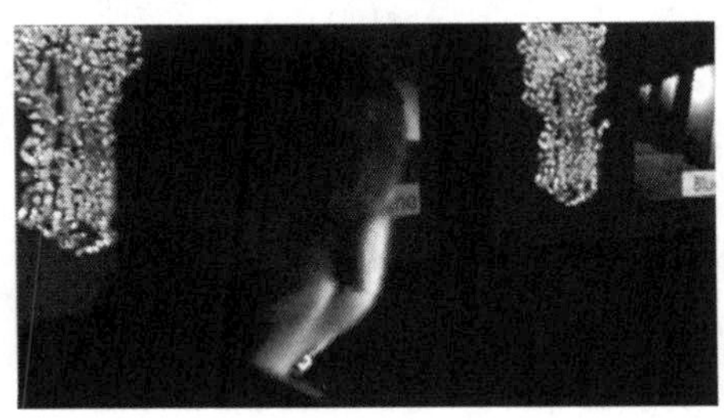

Figure 1.5. IBM promises to transform the planet someday into an object of computation and calculation to forestall future pandemics. This project entails generating detailed models of the world, bird migration pathways, global highway patterns, and flight traffic information. Screen captures, *Deadly Migration: Outsmarting the Avian Flu Pandemic,* IBM Films, 2008.

from mere storytelling. Analysts who make up such scenes insist that their imaginings of possible futures are the products of "rigorous, disciplined, and systematic training."[21] Viewers are walked through this systematic and disciplined mode of thinking as they watch *Deadly Migration*. These narrative strategies of scenario building play a central role in preemptive biopreparedness because they train subjects to consent to treating fiction as reality.[22] Accordingly, these what-if scenarios operate in the documentary to immerse viewers in the conditions of emergence. *Deadly Migration* animates the position taken by U.S. health security experts, calling on viewers to "live in a state of constant readiness" for an unknowable but encroaching future catastrophe, while orienting them toward a hypothetical future through an affect of uncertainty.[23]

In framing future catastrophes, preemptive biopreparedness opens the unknown to quantification, calculations, and measurement, making it a site of technological flourishing. Yet all of IBM's calculated outcomes and computer-modeled viruses are restored to their ultimate precariousness; if the doomsday scenario unfurls through visual and aural overload, it ends, fleetingly, on a frame with no maps, images, information, or numbers. In their absence, a black screen echoes only with a spectral avian call. Because viral emergence turns on the logic of microbial shifting, and because that movement is an endless process that charges the microbial future with endless catastrophic potential, the long-term prospects of avian flu can only appear as an empty placeholder for a time to come that cannot be represented. Exemplifying the belief held by American government scientists and public health officials that we will be "threatened ad infinitum, by new and mutant microbes," the concept of emergence in *Deadly Migration* posits a certain endlessness by reinventing the future threat over and over again, "always already" close at hand.[24]

Media Ecologies and Networked Uncertainty

Fragile, uncertain, and profoundly contingent: how is the message of preemption and the affect of uncertainty conveyed by *Deadly Migration* embedded in the everyday lives of ordinary people in the United States? Both are interwoven into a larger experiential fabric that reveals a host of seemingly disparate actors, agents, technologies, and spaces to be part of the same affective network. Lakoff and Masco claim that such representations of possible future catastrophes work to emotionally manage experts and policy makers by keeping them constantly on alert and perpetually afraid.[25] The larger project of preemptive biopreparedness governs the civil sphere through affect, absorbing a broad nationalized public into an unending and diffuse war on the microbe. *Deadly Migration* summons individuals to take part in a campaign and become "integrally paramilitary, in operative continuity with war powers, on a continuum with them, suffused with battle potential, even in peace"[26] (or in the case at hand, even in the absence of a manifest pathogenic threat).[27] The documentary's representation of mutant viral futures reveals the contours of this discourse of preparedness and preemption, yet representation alone cannot be responsible for the transformation of a "structure of feeling" across broader publics.[28] Beyond representation, *Deadly Migration* is also a material mediation operating through larger media ecologies that constitute the experiential infrastructures of our daily worlds.[29] As Lisa Parks argues, this material "arrangement of audio-visions with other media, architecture, transportation, science and technology, the organization of work and time"—and our encounters with and within these ecologies—generates "experiences, sensation and structures of feeling."[30] Examining the media ecologies of *Deadly Migration* thus brings into view its operational arena.

Deadly Migration's message of preemptive biopreparedness is transmitted in, through, and between webs of people, objects, and spaces—for example, the people, objects, and spaces in the cabins of American Airlines flights in which the documentary was shown. Belted into our seats at an inescapable height of 43,000 feet, we are reminded that the cabin is a site for a range of translocal encounters by the accumulated smudges of fingers on our in-flight entertainment touchscreen from countless strangers from innumerable places. At the same time, our awareness of the cabin's recycled airflow makes us question: "Sharing this air with *whom?*" Later, as we disembark at the airport—that global node of microbial exchange—we pass a poster reminding us how to cough correctly, while an automated sneezing etiquette announcement plays over the terminal speakers. As we return to our daily lives, embedded once more in familiar routines, the message of future pandemic threat engages discreetly, perhaps unconsciously, with a larger media ecology. From *Deadly Migration* and the inflight entertainment system as a viewing experience, from the long hours in the cramped cabin, through the impatient trudge to customs at the airport, the message of preemption implicates and draws together fictional films, flu information tweets, mobile apps that alert us to "outbreaks near me," hand-washing instructions above public sinks, the CDC's online flu calculator, hand-sanitizing pumps in everyday spaces, and the flu vaccine poster at the local drugstore (Figure 1.6).[31]

Coaxing the larger fabric of *Deadly Migration*'s media ecologies into visibility also makes visible its new modes, spaces, scales, and objects of governance. Media become potent affective sites that "premediate"—that is, they make the possible traumas of hypothetical future scenarios a backdrop of daily life—to guide public action and sentiment in the present.[32] These media ecologies

a

Figure 1.6. (a) Touchscreens in an airplane cabin; (b) person using a hand-sanitizing station; (c) coughing etiquette poster in a mall; (d and e) Health Map alerts users to nearby outbreaks in real-time. Author's photographs.

TREAT
here to help.

b

Cover
your

c

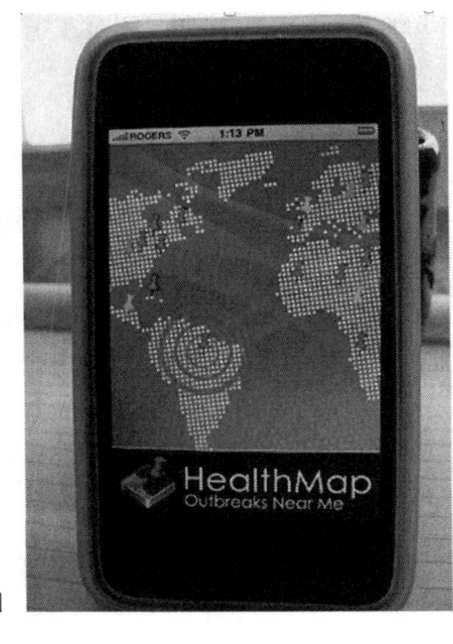
ROGERS
1:13 PM
HealthMap
Outbreaks Near Me

d

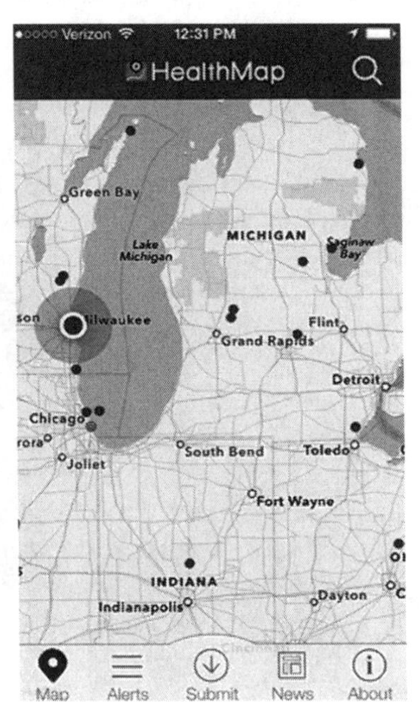
Verizon
12:31 PM
HealthMap
Green Bay
Lake Michigan
MICHIGAN
Saginaw Bay
Milwaukee
Flint
Grand Rapids
Detroit
Chicago
Joliet
South Bend
Toledo
Fort Wayne
INDIANA
Indianapolis
Dayton
Map
Alerts
Submit
News
About

e

orient collective affect toward the future by sensorially immersing their publics in its projected horizons. They make the not-yet-existent biological catastrophe manifest as a medium to be lived through, immersing people in that future before it materializes.[33] Moreover, by transmitting the message of preemption and the affect of uncertainty through the broader world, these media ecologies communicate a sense of precariousness across quotidian sites, spaces, temporalities, and objects. *Deadly Migration* stretches the message of the always-present future threat in and across a broader social landscape, so that an affect of anticipation is threaded seamlessly into the fabric of everyday life.

Atmospheres of Catastrophe

After the wave of posters prescribing an etiquette of coughing, sneezing, and hand washing, after the early flu alerts on Twitter and the novel pandemic projection maps that first appeared in the early 2000s, who really paid attention to such things in the ensuing years, months, or even weeks? They became so embedded in daily life that they soon slipped inconspicuously into the background of quotidian routines. Yet the media ecologies of preemptive biopreparedness still did their work, because they rely as much on inattention as they do on moments of attention. Michael Billig makes an analogous claim in relation to what he terms "nation-making":[34] Nations, once built, require a continuous process of making and remaking. Beyond the overtly "waved flags" of patriotic ceremonies, the ritual elements of presidential inauguration speeches, and other exceptional events, the *sense* of nationhood still depends for its continued existence on "a collective amnesia to mindlessly, rather than mindfully, recall the nation routinely."[35] Billig calls this maintenance of nationhood through repetition "banal nationalism." A nation flags itself daily precisely through forgetting, through the barely noticed encounters with collective signs and symbols: the limp, sad stars and stripes at the strip mall; the red, white, and blue postal truck that passes each day; the same colors on the drooping bunting at used car dealerships. These are the "unwaved flags" of banal nationalism.[36]

In a similar vein, rather than being relayed by representational pyrotechnics, the media ecologies of preemptive biopreparedness rely on their near imperceptibility. The message of preemption is sedimented in the contemporary imaginary "not *despite* its forgettability, but *because* of it."[37] So available do media ecologies make the messages of preemptive biopreparedness that they continue to work even in individual acts of distracted forgetting. When a pandemic scare emerges—as with SARS in 2003 or the swine flu in 2009—face masks, radio and newspaper reports, pandemic watches, and social media light up with activity and flood domestic, work, and leisure spaces. Then, in the quiet aftermath on any run-of-the-mill day, the morning radio might report a flurry of flu activity in a distant part of the world and the newspaper offers a small sidebar on the latest flu-related deaths over breakfast, while the coughing instruction poster still hangs unnoticed in the office elevator, as it has for years. Later, a website might flash an ad for *Deadly Migration* at the top of the screen, while the standard weekly tweet from the Centers for Disease Control goes unseen: "Influenza activity has remained at approximately the same levels as last week."[38]

The sheer ubiquity of such media ecologies has transformed the psychosocial space of everyday life in the United States. Seemingly insignificant, these elements are in fact integral and material components of the mediated environments of contemporary public health that effectively keep collective senses trained on impending catastrophe. *Deadly Migration* operates in a larger landscape of messages, warnings, and reminders that constitute a "national theater"[39] in which people have no choice but to contemplate that catastrophe daily, if distractedly. The work that these media ecologies do is only reinforced when they fail to predict with substantial accuracy. The limits of various techniques to apprehend risk, their very ubiquity and the *constancy of their failure,* are an essential part of the affective logic of this national theater. In other words, the perpetual failure to apprehend risk adequately is embedded in the project of preemptive biopreparedness: it ensures that even if the fateful pandemic does arrive (as with SARS in 2003, swine flu in 2009, and Covid-19 in 2019), the target slips through our grasp, thus requiring people to reinforce their commitment to the continuance and building of the project. Emergence discourse exhorts us to forever ready ourselves against the "next one." It makes threat present everywhere, while suspending people under a future of unyielding pandemic possibilities, a kind of affect Brian Massumi describes as "suffus[ing] the atmosphere."[40] In this way, such media ecologies evoke a perpetual, low-grade state of crisis that I term "atmospheres of catastrophe."

Better Safe Than Sorry

The everydayness of sensorial immersion in conditions of ongoing crisis and the shaping of quotidian experiences across an affect of uncertainty are characteristic of what Lauren Berlant calls "crisis ordinariness."[41] In crisis ordinariness, crisis merely and undramatically saturates the fabric of everyday life and is felt no more concretely than as "something in the air." By transmitting the message of ever-present, ever-potential pandemic risk so ubiquitously and diffusely, the media ecologies of preemptive biopreparedness transform the experiential fabric of everyday life, making and remaking the nation-space as a domain of lived, ordinary crisis. And because the conditions of quotidian crisis register affectively before they manifest in other forms, responses to living in, through, and against crisis are legible not in marked gestures, overt beliefs, or in any other more tangible outcome. Instead, those responses appear discreetly, precisely because ordinary crisis is "suspended in the sensorium without hitting a nerve."[42] Our affective relation to crisis congeals into undramatic coping responses, which may include half-noticed or vaguely recalled thoughts, or new habits and modes of existence as we adjust to our conditions of pervasive and systemic crisis. Consider, for example, reader responses to news about chilling claims that the next pandemic will be the end of humanity as we know it. One reader, under the moniker of "godfirstcountrysecond," comments on CBS's article "Pandemic Influenza" to dismiss the frenzy over the pandemic potential of the 2009 swine flu episode as fear-mongering hype, but makes a notable admission: "I'm starting to think about buying facemasks for the whole family, but the media has run amok with this."[43] David g. and SSW, two separate reader-commenters on an ABC article on the same topic, caution: "Plan for the worst," and "maybe store some food and water, just in case."[44] Consider, as well, the ease with which we might share the sentiments of individuals such as Jen Lamas, a Facebook commenter on an ABC story about flu-related deaths. They urge the online public to get the seasonal flu vaccine regardless of efficacy: "Regardless if the strain is correct or not, better be safe than sorry!"[45]

Better be safe than sorry. We can understand such "just in case" statements, and their addled gestures, as instances of what Berlant calls "genre flailing."[46] This concept helps make sense of affective confusion: "Genre flailing is a mode of crisis management that arises after an object, or object world, becomes disturbed in a way that intrudes into one's confidence about how to move in it." We genre flail "so that we don't fall through the crack of heightened affective noise" into states of chaos, despair, or psychosis. We improvise "like crazy" in an almost literal sense. Genre flailing is often expressed in fruitless and unimaginative movements. When crisis enters into a phase of ordinariness, flailing—"throwing language and gestures and policies and interpretations at a thing to make it slow or to make it stop—can be tediously dull and unimaginative: a litany of lists of things to do, to pay attention to, to say, to stop saying, or to discipline and sanction." The "just in case" comments are flailing attempts to grasp whatever they can when "What is happening?" comes to dominate the general tone of existence.[47]

Berlant notes that, under the pressure of ordinary crisis, we often look to "normal science" or "common sense" as a type of established clarity to stabilize a disturbed world. At the same time, in the context of emerging microbes, we are forced to see that these points of affective refuge can no longer offer the same reassurance. Instead, the instability of the world is once again confirmed as all manner of technoscientific interventions fail to proffer something true and firm. When we cannot think of what to say or do, whatever we do is "off the cuff," or despite, or at the last minute, or "just in case": all instances of genre flailing. On a personal note, I too have been pressed to genre flail. During the present pandemic, I too have carried out various measures to avert microbial infections and take care of myself and others. During the successive waves of Covid-19, I have adopted measures and adjusted my being in the world in ways that have seemed both illogical and logical. And I do these things out of a sense of fear and desperation because the expert systems—Berlant's "normal science" and "common sense"—are themselves flawed and uncertain. When things feel out of control, we habitually hold onto testing systems, healthcare networks, biomedical enterprises, supply chains, and governments. Yet were those who drank bleach, consumed Ivermectin, or took antimalarial drugs merely "stupid people doing stupid things?" Or were those gestures also instances of "improvising like crazy?" Or, again, was it unrealistic to wear masks to enter a restaurant only to shed them over the course of a meal? We can read these conflicted gestures instead as aporetic flailing at a time when "we're all disoriented, or in crisis and wanting to fix the world," but have no idea where and how to begin, or how to get out.[48]

Better be safe than sorry. These "just in case" sentiments point to a politics of trust and confidence at play when the very substances of relation—touch, proximity, contact, breath, and the ether between every possible being—are turned inside out as substances of disconnection. In the anonymous and impersonal worlds embodied by our thoroughly networked realms, proximity in time and space thickens a sense of personal relation. This nearness, Anthony Giddens argues, generates a specific type of knowledge.[49] Proximity, he suggests, enables us to have a sense of certitude about the day-to-day because it establishes a sense of *personal* bonds in a world that would otherwise feel "phantasmagoric."[50] Contrary to Giddens, under the logic of preemptive biopreparedness, I argue that proximity and contact no longer sustain relations of confidence. Instead, in the ordinary crisis of preemptive biopreparedness, proximity and contact are constant reminders that every entity is a substance of microbial entanglement and possible exposure, as is air itself. The constancy of crisis, our immersion in the ubiquity of possible catastrophic scenarios, reveals Giddens's "fateful moments"—breakdowns in confidence in our day-to-day lives, our sudden awareness that contemporary systems are ultimately unmanageable—as ongoing conditions of life in pandemic modernity.[51] What more potent sign that an air of mistrust permeates proximity and contact (the very relations that purportedly form the basis for restoring "confidence" in daily life) than the surge of "touchless" quotidian spaces such as public bathrooms, soap dispensers, and shopping malls? The "just in case" appears in the hesitation to eat Chinese food during the 2003 SARS pandemic, in the fleeting and uneasy tickle in the back of the mind when reaching for a magazine in the lobby of the doctor's office, in the mindless awareness of what we may be touching when grasping subway poles, in the now unthinking habit of pumping hand sanitizers to punctuate space and time in the course of daily tasks.

These seemingly insignificant acts and thoughts, and their gradual insinuation into worlds of habit and feeling, are congealed in and through media ecologies even in the absence of a belief in their message. Such responses mark out emergent modes of living as people attempt to carve out ways of carrying on while stuck in a holding pattern of uncertainty. The meaning of any particular message, Michael Schudson suggests, does not always work by convincing people to believe it. Instead, messages operate by articulating a "way of experiencing," rendering that structure of feeling or thinking "so available that they manage to bring some images and expressions quickly to mind while making others relatively unavailable." The sheer proliferation and utter ubiquity of such messages function like "switchmen on the line of history." In a world full of competing ways of thinking and acting, they "bring one way of experiencing above all others closer to mind."[52]

"Probably," "maybe," or "I try not to think about it," or "I don't really believe, but just in case": these responses reveal the extent to which uncertainty itself has become one of the most readily available frameworks through which people experience nationhood. Like Billig's banal nationalism, the media ecologies of *Deadly Migration* flag the nation on a daily basis as a space suffused with microbial risk that demands a perpetual state of caution. Such media ecologies neither help to manage risk nor mitigate it. Instead, they transform our lived relation to risk in such a way that we literally learn how to live *with* it—as Schudson might say, "we get used to it," or more accurately, "we get used to not getting used to it."[53] Learning to live with risk leaves us to wonder, in the background, whether invisible and life-threatening forces are intruding into our air, campaigning across nearby surfaces, lodging in our bodies to bring with them fever, chills, and occasionally death.

I do not mean to suggest that everyone is afraid of dying by flu, nor that there can be a homogeneous response to living with risk in the ways described above. But the affective responses noted above do register the kinds of impasses and mundane habits that seep into routinized modes of existence in the space-time of everyday crisis. When, for example, did I start to sneeze into the crook of my elbow? It is impossible to pinpoint this in time, as it is now the norm in North America. If we reflect on the now-habitual practice of hand sanitizing—a habit once deemed to be the disorderly behavior of the obsessive-compulsive—we might see in it a kernel of the unthinkable, uncertain logic of *Deadly Migration*'s "what-if scenario." In these everyday responses to perpetual pandemic possibility, we might also see a national security discourse domesticated in our hands.

One Nation under God Knows What

If the media ecologies of *Deadly Migration* suffuse the nation-space with the charge of everyday crisis, how does its force translate into the affective conditions of a national collectivity? How to move from the world of individual habit and thought to the emergent mode of nationhood that appears under the project of preemptive biopreparedness? Rather than noting the peculiarities of individual experience, the affective responses outlined above trace a structure of feeling through which pandemic modes of nationhood emerge. As Berlant argues, while such intimate, unthinking, and seemingly insignificant expressions render legible the shifting modes, habits, and attitudes of *individuals* learning to live in states of uncertainty, they also mark out the tone of *collective* experiences exemplary of a *shared* historical time.[54] Media ecologies of preemptive biopreparedness compose affective atmospheres as "a class of experience that occurs before and alongside the formation of subjectivity, across human and nonhuman materialities, and in-between subject-object distinctions."[55] In this way, they render collective affect infrastructural to contemporary national life. Masco makes a similar argument about post–Cold War America, arguing that ordinary emotions of anxiety have coordinated citizens as members of the U.S. national security state such that anticipatory fear is now "coded into social life."[56] In our context, these media ecologies place uncertainty at the center of American governance in ways that manage the affect of the civil sphere. They are perpetual exposures to *Deadly Migration*'s core message that we are always only "one step away from a deadly pandemic," and thus that "the future is ever uncertain, because unimagined new diseases surely lie in wait, ready to emerge unexpectedly" even twenty years after concerted efforts to counter their threat (2012 publication by the National Institutes of Allergies and Infectious Diseases). For this reason, "offense is the best defense" against viruses (2013 Lysol brand slogan), because it is "better to be safe than sorry" (2004 US IOM publication).[57] The efficacy of preemptive biopreparedness cannot be gauged by the certainty by which one makes decisions in response to infectious disease threats. Rather, in line with Berlant, its impact might best be measured in these moments of doubt and impasse—in our general inability to act with any certainty, but to flail in any case.

While most people may not actually fear death by flu, *how* individuals fear as a *collective* through the shared affect of precarity reveals the insecurity at the heart of this nation-making project. Pandemic nationhood is based on a slow erosion of confidence in the day-to-day, forged through the loyalty of our senses to the message of preemptive biopreparedness. The mode of collectivity that emerges results in social relationships permeated with anxiety. This sense of nationhood is grounded in the disquieting recognition that disentanglement is unachievable, that "their air" is inseparable from "my air," that "exotic" pathogens from "distant" lands are embedded in our most quotidian encounters, and that their dangerous potential is charged within every register in our lives. And the personal UV air filters worn around the collar, face masks in public spaces, or hand sanitizers toggled to bags can be seen as parts of the affective ecologies of a nation organized under ordinary uncertainty. As with the "just in case" moments examined earlier, these moments mark textures of hesitation and doubt inherent in the fabric of ongoing crisis. They can be seen as instances of a collective withdrawal from the shared and sharing substance of air that Peter Sloterdijk and others describe as the very stuff and substance of our becoming-together.[58] In the course of the hyperhygienic campaign, what occurs instead is a withdrawing-together from the conditions of our material interrelation. In its wake, the national body politic scans itself as potential stranger, foreign agent, and incipient enemy.[59] This misrecognized self that attacks itself as stranger is in line with the conditions of autoimmune disorder that Roberto Esposito characterizes as a "refusal from commonality"[60] on a national scale. As a refusal of commonality, a pathogenically made nation does not produce the kind of nationhood invoked by Benedict Anderson, in which a "broad horizontal comradeship" is fashioned through a common awareness of shared experiences, histories, and cultures to unite a band of otherwise disparate strangers within a conscribed geographic space.[61] What it produces instead is a sense of nationhood rooted in social fracture.

This is the nation to which our airline traveler returns. A place where shaking hands, going bowling, pork, eggs, Chinatown and "the Chinese," classmates at their child's school, neighbors who do not get the flu vaccine, grocery cart handles, and door knobs are all enveloped with suspicion. This person is at home in a place of everyday displacement, a realm of uncertainty about the people, localities, spaces, and things within the pathogenic nation-state. Their ordinary anxiety, like the unease of others around them, is a fear of contact and proximity itself. Indeed, while the CDC and Department of Homeland Security together assessed the legal boundaries of social distancing measures in the Social Distancing Law Project, the CDC's three-to-six-foot recommendations began to take root on the ground from 2009 onward.[62] The recommendation to maintain that distance between individuals during peak influenza periods once sparked National Public Radio's tongue-in-cheek suggestions for alternate handsless ways of greeting one another (Figure 1.7).

While NPR's proposals were made in jest, they nonetheless highlight the uncertainty fashioned through the mediation of the pervasive, never-ending threat of a deadly pandemic. Rather than being "one nation, under God, indivisible, with liberty and justice for all," the space and logic of preemptive biopreparedness works to unite "everyday Americans" as one nation bound by "God knows what." It calls on individuals to forever ready themselves in the

The Wave

A cowboy howdy or the royal wave are good alternatives to the classic handshake, especially in the colder, drier months of the year when the flu virus stays around longer after you cough or sneeze. That means more virus in the air to inhale... and more chances for you to pick it up from someone's hands.

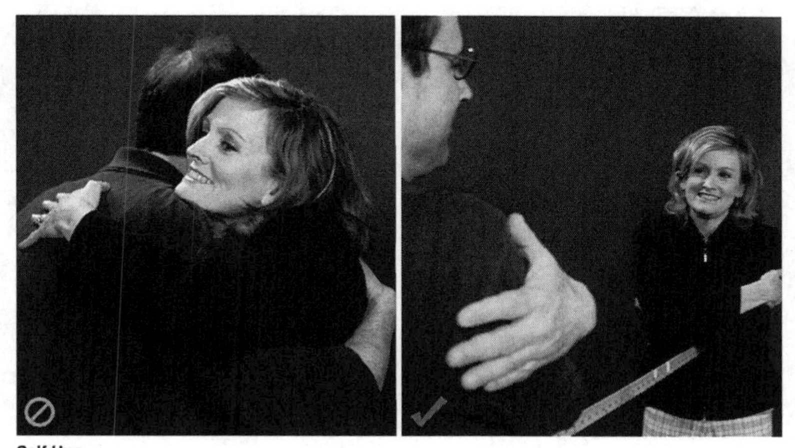

Self Hug

Hugging yourself might seem odd at first, but surely it can catch on... maybe.

Figure 1.7. National Public Radio offers suggestions for how to greet one another while keeping a distance of 3–6 feet. National Public Radio, 2009.

face of unending and unknowable catastrophic biological futures. We may join hands, but in the touchless era of emerging infections, only metaphorically, in what Ulrich Beck terms "a solidarity around anxiety":[63] a low-grade fear of the very things that compose and decompose the contemporary American nation-state.

Pathogenic nation making also forms the building blocks of a U.S. post–Cold War global project. Perhaps most obviously, the discourse of emerging microbes gave rise to a particular U.S.-led formation of global health. But this new project of global health also called forth new forms of international cooperation, provided new ways of apprehending ecosystems and environments, reshaped epistemological frameworks, and took hold of objects, processes, and life itself across open spaces and times of microbial emergence in ways that reverberate still. How does it matter that this nation-making project became the basis for global remaking on multiple scales? In what ways has this ordinary uncertainty become a critical affective basis on which life, knowledge, politics, and global systems are remade? I probe such problems across the following chapters to trace how the project of microbial resolution remakes the world in the image of its perilous instability. In the next chapter, I return to the 1990s to consider how the charged world analyzed in this chapter came to prominence in the first place.

2 Materializing Emerging Microbes

It is August 1997. Above the waters of the Bering Strait in Northern Alaska, a small aircraft crosses a broad expanse of permafrost and lands near the Brevig Mission in the Inupiat territory of Sitsaq/Sinaurag.[1] Retired pathologist Johan Hultin is aboard for his second trip to the site. His first trip was in the early 1950s, when he was a young PhD student at the University of Iowa. At a campus cocktail party, Hultin heard a visiting professor mention the high likelihood that the 1918 influenza virus had been cryogenically preserved in the bodies of the Inupiat people who had died from the disease,[2] seventy-two of eighty inhabitants over the course of five days (Figure 2.1). Travelling to the Brevig Mission settlement in 1952, Hultin and three colleagues exhumed some of those bodies from the mass grave and extracted lung tissue from four that exhibited signs of pulmonary hemorrhaging, a well-documented effect of the 1918 flu.[3] They brought the specimens back to Iowa with the intention of reviving and studying the virus in the university's facilities, but the samples had materially degraded over the course of the journey.[4]

In 1997, Hultin decided to embark on a second trip to the Brevig Mission after perusing an article in *Science* about the work of Dr. Jeffery Taubenberger and Dr. Ann Reid, two molecular biologists at the U.S. Armed Forces Institute of Pathology (AFIP).[5] Taubenberger and Reid hoped to reconstruct the 1918 virus, unlocking the dangerous "secrets" encoded in its genes to prepare for the possibility of future influenza pandemics.[6] The field of molecular biology developed rapidly throughout the 1990s; the invention of high throughput genetic sequencing technologies, as well as

Figure 2.1. Gravesite containing the remains of the seventy-two of the Inupiat who died of the 1918 flu, Sitaisaq/Sinauraq, Brevig Mission. Alaska. Image courtesy of Angie Busch Alston, thealaskateacher.com.

methodological developments in polymerase chain reaction, made it possible for scientists to reconstruct the entire genetic sequence of a virus and then remake viruses from partial samples. But the AFIP team was working with badly compromised specimens: lung tissue preserved in paraffin wax that had damaged the material integrity of the virus. From this imperfect source material, the team managed to reconstruct one of the eight gene sequences of the 1918 virus's RNA.[7] Hultin contacted Taubenberger to tell him about the bodies he had found preserved in the frozen tundra when he was a graduate student, and offered to make another journey to Alaska to obtain new specimens for their research. Taubenberger agreed, and Hultin quickly organized the trip from his home in San Francisco.[8] Withdrawing $3,200 from a personal savings account to finance the mission, he set out on the second expedition one week later.[9]

At the same mass grave in northern Alaska, Hultin excavated once again. Assisted by several Inupiat villagers, he found the body of a woman whom he named "Lucy." Chipping away at dense layers of frozen fat tissue, Hultin found a pair of pristinely preserved lungs.[10] He sent his findings, this time efficiently protected, to the

AFIP headquarters in Washington, D.C. Using Hultin's samples, Reid and Taubenberger's team was able to sequence the 1918 gene in its entirety.[11] Viral RNA is the genetic blueprint containing the plans and instructions for new viruses. With its complete RNA reconstructed, the 1918 virus could in principle be brought back to life by creating plasmids, tiny circular pieces of DNA for each of the eight segments of the gene. AFIP sent the RNA sequence to Dr. Peter Palese, the molecular biologist who had pioneered the creation of plasmids. Once his own intervention was complete, Palese forwarded the specimens to Dr. Thomas Tumpey at the U.S. Centers for Disease Control. Working in the Department of Defense's level-3 biosecurity laboratory, Tumpey used reverse genetics to insert the plasmids into human kidney cells, which they could instruct to reconstruct the RNA of the complete sequence. Tumpey and his colleagues watched the cultured cells for weeks to see if they would produce the 1918 virus. On the day the virus first appeared in his cell cultures, Tumpey—perhaps anticipating the significance of this moment for the unknown frontiers of microbial emergence—sent a one-line email to his colleagues, quoting the celebrated words of Neil Armstrong as he walked on the surface of the moon for the first time: "That's one small step for man, one giant leap for humankind."[12]

This chapter tracks how the immaterial futures of emerging microbes were materialized as a scientific object, a knowledge project, and a paradoxical force to be secured throughout the world. I trace the development of the emerging microbes concept in and after the 1990s, showing how their materialization was structured by the political and historical context of the post–Cold War period. I examine three defining and disparate threads of that historical moment: the genetic revolution, the end of the Cold War, and the development of a new paradigm of global health, through the lens of emerging microbes. My analysis holds these three strands together to underline the close, interlinking between the global campaign against the worldwide threat of emerging microbes, U.S. national self-refashioning through the microbial realm, and the new ecological configurations of post–Cold War conflicts. This narrative of the 1918 virus joins the three core concerns of this chapter. First, I provide an account of early research into emerging microbes as a process contingent on methods and techniques of materialization specific to the period. The project of microbial resolution sought to transfigure the discursive futures of emerging microbes into living material in the world. I situate AFIP's experiments in a historical context when developments in the evolutionary and genetic sciences come together with information discourse to reset the epistemological and material boundaries of scientific inquiry. These new boundaries produced a conceptual understanding of emerging microbes as containing vast and endless frontiers that teem with possible futures. Although Tumpey may have used Armstrong's words with a certain light-heartedness, the interstellar and microbial voyagers yoked together in his quotational quip may have more in common than he intended. Such tropes reveal how research into emerging microbes was cast as an exploration through uncharted times, spaces, and scales. AFIP's research to plumb the futures of emerging microbes parallels other scientific pursuits to draw frontier spaces—the oceans, the Antarctic, and outer space—within the grasp of scientific knowledge. In highlighting this exploration narrative, my intent is not to replicate the sequencing of the 1918 virus as a heroic origin story with a central cast of (white, American) characters; I take up the problematics of colonialism and biocolonialism in chapters 4 and 5, but raise the narrative frame here to highlight the scientific production of emerging microbes as entities containing inexhaustible frontiers.

My second concern in this chapter is the formation of emerging microbes as objects in the historical context of post–Cold War science and geopolitics. The absence of war plunged the United States into a multifaceted crisis without an obvious external enemy (such as the Soviet Bloc) against which to define itself. In that vacuum, emerging microbes became positioned as a novel kind of national foe that reshaped the nature of global threat and conflict after the Cold War. The paradoxical material status of emerging microbes as potentially everywhere but actually nowhere was taken up in a new American security paradigm at the end of the twentieth century. I examine how this new logic relied on retraining the sensorial horizons of vision and mediation to be constantly attuned to this perpetual and elusive opponent.

The 1918 retrieval narrative also reveals how the project of microbial resolution necessarily exceeds the space of the laboratory and the worlds of technoscience: it engages the elemental world of permafrost, oceans, and air; it moves across territories; it draws upon global systems and infrastructures. The third concern of this chapter is to make legible how the project of microbial resolution extended warfare against the ubiquitous threat of microbes across the open spatio-temporal continuums of emergence, showing how the war against not-yet-existing microbes took hold on a planetary scale. Studies of pandemics often attend to the ways in which anxieties around zoonosis (the process of species-jumping viruses "spilling over" from animals to humans) draw a range of biological taxa into complex regimes of governance in an effort to "secure life."[13] This focus on species obscures what I argue is a far broader footprint of the U.S.-led war against emerging microbes; its biosecurity practices in fact confound the assumption of animal/human dichotomies, and no longer hold life/bios as a primary object. I shift away from a focus on managing life across species to make legible how the war on microbial emergence apprehends a world utterly saturated with risk, made up of both entangled natural and manufactured systems and living and nonliving things—what I term the "transnatural biosphere." Tracing these cross-scalar (microbial to global) and transmedial (bodies, infrastructure, oceans, atmospheres, supply chains) dimensions reveals how the project of microbial resolution imagines the transnatural biosphere as its battleground. I question the critical scaling effect of this war, asking how the United States worked within the realms of the ill-defined, the hypothetical, and molecular to renew its footing in a post–Cold War world, while saturating that world with its new security logics. I examine the aesthetic and affective tailings of this new global project structured by the imperative to resolve the irresolvable material paradoxes of emergence.

The Multiplying Edges of Emergence

Reid and Taubenberger peer at a sheet of film cradled between them (Figure 2.2). The laboratory in which they stand is a cluttered space of grey and beige, bulky equipment, wire prongs, cramped

Figure 2.2. Taubenberger and Reid look at the genetic sequencing image of the 1918 virus. MIS 377212. National Museum of Health and Medicine.

shelves, ancient sticky notes, and yellowing photocopied signs. But their backs are turned to all that clutter, and their eyes are fixed on the series of marks that reveals the initial segment of the genetic sequence of the 1918 virus. They stare into the rendered gene spellings as if searching for significance; by elucidating this genetic code, the two molecular biologists hope to make legible the futures of emergence.[14] Their bodies are planted in a familiar world, but their gazes seem to extend to a time and place beyond the monotony of their quotidian surroundings. They almost appear to wish to enter the film, to move directly into the molecular realm itself. The glints of light flashing off the two rings and glasses triangulate a vision structured by the desire to voyage into the molecular realm in search of their elusive epistemological object.

Reid and Taubenberger are performing this work and this scene at the height of the genetic revolution when new technologies allowed scientists to peer into the molecular registers of life, conditioning new ways of seeing and investing the scientific gaze with new desires.[15] The structuring belief and desire of the genetic revolution is that seeing the molecular allows us to read life beyond matter: to read it as transparent information. Whereas microscopes or telescopes enlarge or produce more proximate images of objects, viewing on a molecular scale offers a glimpse beyond the surfaces of materiality. Colin Milburn suggests that technologies of seeing on molecular scales—part of what he calls "nanovision"—operate through a productive relation between blindness and sight. Nanovision presents the molecular realm as existing beyond our visible grasp while simultaneously opening up the inside of molecules as an immaterial "elsewhere," a limitless frontier space that can be entered and explored via vision.[16] But what is visible beyond matter? Milburn suggests that nanovision is formless, a kind of amorphous goo and splatter.[17] More accurately, for those working in the genetic sciences, to see beyond raw materiality is to see into the realm of information. Matter at one end of the ontological spectrum, information at the other: the genetic revolution is built on a stance that cleaves the two as separate and distinct. This view posits information as a preexisting structure that is the essence of matter, entertaining the fantasy that to see genes is to see information itself: transparent, pure, and available. Information is in turn configured as a peculiar kind of object, an essential extract of life that is both the memory of its form and its logos.[18]

Yet Lily Kay cautions against reading genes as the codes of life, because this metaphor took hold for historically contingent reasons.[19] Such metaphors, such scriptural terms, such tropes of writing, such figures of code, data, and information are at once illuminating and obfuscating for molecular biologists. "Information" does not describe what genes actually *are.* Prior to their framing as information-forms, genes were primarily referred to in terms of *specificity,* a designation used to address genetic constitution as a series of material correlations between the structure of nucleic acid and the structure of protein. When we speak metaphorically of genes as code or as elements within "the book of life," we're referencing the same material mechanism. However, these dematerializing and textual metaphors (including information, code, communication, language, syntax, transcription, grammar, meaning, and programming) became firmly installed within the field of molecular biology as the dominant mode of describing genetic ontology.[20] Yet despite its inaccuracy, Kay demonstrates that the information metaphor became solidified within molecular biology precisely because it aligned that discipline with a culture deeply invested in cybernetics, linguistics, cryptanalysis, and electronic computation, all profoundly shaped by postwar culture. Molecular biology, in its genetic phase, coalesced within a Cold War paradigm of communicational and computational models of understanding.[21] And the terms used to describe genes are situated and shaped within that discourse, which includes a preoccupation with surveillance and the interception of hidden messages.

This textual understanding of genes underwrites the conviction that scientists may decipher the futures of emerging microbes once the secret script contained in their genes is revealed. In keeping with their Cold War roots, genetic technologies work as decryption operations to render genetic codes legible. Reid and Taubenberger are working within this framework, but tasks like genetic decoding and cryptanalysis are further complicated by two important and interlocked scientific developments. The first is that genetic matter is not static. Microbes constantly transfer genetic matter among themselves, which is imprinted in microbial RNA/DNA. This genetic matter descends vertically, from an "origin" to subsequent generations, which led scientists to assume a stable genealogical lineage. In the second half of the twentieth century, however, scientists discovered that microbes also exchange genetic matter horizontally: that is, with their contemporaries, and their development is shaped by ambient and external conditions (the material contexts of their existence, such as the rise of antibiotic use).[22] Microbes are what Hannah Landecker describes as "a pastiche of other slowly accrued, vertically inherited features."[23] As "historical composites," they form "multifarious" assemblages that defy clean, genetic (and perhaps even ontological) categorization. This development meant that microbes and their genetic information could no longer be understood using linear, stable frameworks of classification. If genetic development was messy and unpredictable, reading the future pathways of microbial emergence would be akin to solving an equation in which the values constantly recalibrate in incalculable ways. The second complication is the shift in the 1980s from a Newtonian world of order to the study of complex systems in scientific theories of development and change.[24] As such studies were devoted to the analysis of disorder, one name for this branch of study was the science of chaos.[25] Here, the evolution of systems (including those of microbes) was thought to arise out of dynamic interconnections born of crisis.[26] Crisis was generative; rather than leading to entropic decline and depletion (as in the homeostatic Newtonian model), injury, stress, and threats to life were seen as productive and important under chaos theory. Evolutionary pressure would push life systems across thresholds necessary for autopoietic creativity. A larger science of "emergence" developed in the natural and social sciences that took the future as an object to be deciphered. But what is there to decipher when complex systems are always in inchoate states of possible transformation? And how to make sense out of the disordered states of an illegibly complex world?

While developmental processes were constantly transforming in disordered ways, the science of chaos does not apprehend the world as absent of order. Rather, it restructures the relation between chaos and order, holding that a hidden, deep structure lies within complex systems. Complex systems are conceptualized as "rich in information rather than deficient in order"; as Katherine Hayles notes, information was seen as the "connective tissue holding the system together,"[27] and its hidden structures could be understood once the patterns underlying the disorder were decoded. Genetics, evolutionary systems, cryptography—fields deeply shaped by Cold War systems thinking—fashion the view that information is both a force animating life and a key resource through which to speculate about its potential futures.[28] U.S. scientists envisioned emerging microbes as objects that position scientific inquiry in a speculative space, and microbial resolution as a Sisyphean task of chasing the unknown futures of unpredictable change. This is the context in which Reid and Taubenberger view the rendered gene spelling of the 1918 virus from a vantage point between the edge of the existent and the immaterial realms of the unknown. I turn now to the next step in Reid and Taubenberger's work, and through it, to some of the ways that figures of emerging microbes and their possible futures were materialized as scientific objects and operable entities in the world.

Signifying and Materializing Emergence

Positioning microbes as endless frontiers to be explored is political as much as it is epistemological. In preparing for all possible microbial futures in advance, the project of countering emerging microbes is aligned with a broader national defense mandate developed by DARPA in the 1960s—to "never again be surprised."[29] This national mission to fend off surprise itself became an important part of an American Cold War rationale to increase technological, scientific, military, and institutional capacity in exploratory and often highly speculative experiments. By porting its security strategies and resources to the unknowable future, the realm of the not-yet was effectively transformed into a generative site for defense building and experimentation. From DARPA's perspective, such efforts would effectively render the United States insusceptible to the unpredictable and, therefore, make it always prepared for anything. The project of microbial resolution works here not necessarily to prevent pandemics, but to produce the incalculable, immaterial, and endless frontiers of microbes as sites where uncertainty itself can be expressed and experimented with. "Future building" seems a fitting term for such efforts to turn the very unknowability of the future into spaces that can be entered, studied, and explored.

How to materialize endless alien frontiers into sites of future building? This common question confronts researchers who face all kinds of vast, unexplored, and unknown frontiers. Stefan Helmreich, for example, examines how marine biologists attempt to bring alien and inaccessible zones of the deep sea closer to human domains of meaning and value through scientific measure.[30] Jessica O'Reilly studies the ways in which researchers apply technocratic measures to similar ends in the context of the "seemingly blank" vistas of the Antarctic.[31] Through such work, those scientists seek to make meaningful worlds out of what Western scientific positivism would regard as the inhuman emptiness of untrodden and "uninhabitable" frontier spaces.[32] Reid and Taubenberger's research on emerging microbes is analogous to these scientific quests to make knowledge out of unknown frontiers. After they sequenced the entire genome of the 1918 virus largely from Hultin's samples,[33] they compared that information with samples they were collecting from various mammals and waterfowl (primarily swine and migratory birds). By comparing viral strains, the AFIP research team was attempting to decipher the protein structures of multiple viruses to determine where the next major pandemic might develop.[34] Their experiments suggested that the 1918 virus, both avian and mammalian, was genetically intermediate, which gave rise to the theory that viruses originate in wild birds, and then mix in mammals (swine) to become human-transmissible. Alongside these valuable insights into the evolutionary behavior of microbes, Reid and Taubenberger's research is also a crucial example of materializing the immateriality of emergent futures as spaces to be worked or experimented upon. When the AFIP team reads the existing genetic spellings of the 1918 flu virus and compares them with contemporary strains, they are not seeking or seeing a history—that is, they are not looking *for* the 1918 virus, nor are they searching for avian or swine flu per se. Although they are studying their specific genetic spellings, they are scanning beyond these samples, beyond even the specific etiology of a given disease or microbe. They are looking to lend structure and meaning to the immaterial and looming futures of microbial life. Their overall interest is not what these microbes once were or where they came from, but what they might become, and how and where they might appear at an indeterminate future point. Emergence itself is the object of their scientific quest.

Unlike scientists who study the genetic origins of microbes by looking backwards to trace a historical relation through vertical inheritance, then, AFIP's efforts look forward. Their research gives contour and definition to the nonknowable realms of emergent microbial futures by grounding the latter in a knowledge of existing diseases. Deploying the familiar to give structure and meaning to the blank and ill formed is a method that Lisa Messeri terms "double exposure."[35] In her ethnographic study of exoplanet researchers, Messeri examines how these scientists prepare for future missions to unknown, interstellar regions. Apollo team members, for example, use the Utah desert as a high-fidelity simulation site for Mars. They test-drive lunar rovers, try out spacesuits, and inhabit dwellings on earth to train for living and working on the alien planet. They deploy the knowledge that they form in a near and familiar space as a proxy through which to grasp and experience a distant and alien one. By overlaying an experience of the familiar onto an unknown cosmic world, scientists transform vast and abstract outer spaces into more definable places that can be entered and explored,[36] thus giving semiotic and imaginable shape to that terra incognita. Similarly, when Taubenberger and Reid peer at the genetic information of existing flu strains to speculate on the potential origins of coming pandemics, they are future-building a vision of microbial emergence through existing diseases. This temporal projection structures the blank horizons of "emergence" through an epistemic semiotics constructed around definable historical examples. This logic also undergirds the National Institutes of Medicine's use of existing and historical examples such as dengue, malaria, and the 1918 influenza to ground the intangibilities of microbial futures. In both instances, the scanning of past or present viruses works as a double exposure; it transforms the conditions in which researchers might envision and give form to the unknowability of emergent life by anchoring it in something familiar. In this way, the emptiness of emerging microbes is rendered dense with meanings, implications, and possibilities.

What meanings coalesce around emergence when framed by the 1918 narrative and contemporary viruses? The semiotic and temporal transpositions of these double exposures create the future as a constant expansion of threat. In order for the uncertainties of emergence to be made into governable objects and experimental sites, they must be materialized. This drive shapes the way U.S. scientists envision and flesh out (in some cases literally) microbial futures: Palese injecting RNA strands into plasmid rings; Tumpey gazing into his petri dish at the ovoid microbes manifesting in his cultured kidney cells; a CDC lab assistant candling the amniotic sac of an egg to assess it for signs of microbial life (Figure 2.3); a researcher aspirating viral particles into test mammals; Taubenberger surveying catalogues of protein spellings to trace the constellations of amino chains that might foretell biological catastrophe; genetic techniques used to engineer the highly human-to-human transmissible avian flu virus discussed in chapter 1. Microbial resolution involves a set of operations and discursive practices through which emerging microbes are transmuted

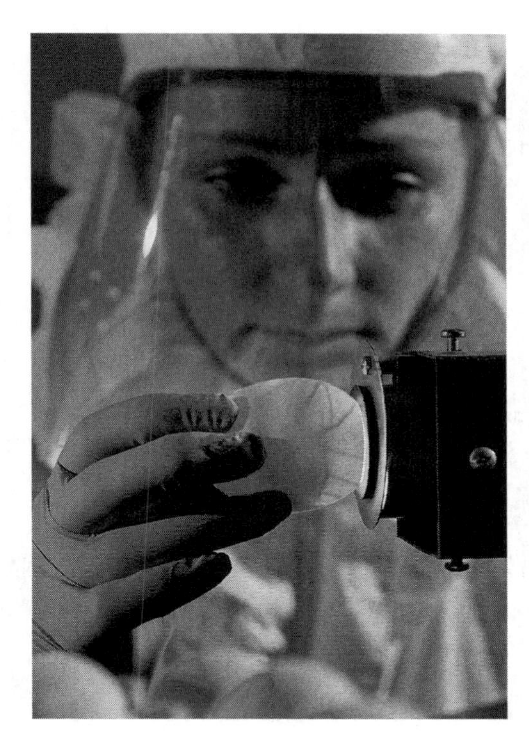

Figure 2.3. A CDC lab technician "candles" an egg to show the embryo within. Photographed by James Gatheny, 2019, image provided by CDC/ Robert Denty, Public Health Image Library.

from immaterial objects to workable matter within the world, often through all-too-familiar tropes of territorial appropriation and control. When scientific positivism encounters realms that extend farther than anyone can foresee, researchers reconfigure them as "extraterritorial space[s] against which public, private and national interests [can then be] projected."[37] The futures of emerging microbes appear as an extraterritory that can be plumbed and mapped, claimed and invested. In this process, a broad range of imperatives linked with the post–Cold War security state are renewed and take root.

Emergence and the New Security Rationale

All that emerging microbes might encompass can be brought within the American security project that surfaced after the Cold War's end. All possible futures, planetary ecologies, novel life-forms, and bioinformation, as well as all places where microbes might be found—in waterways, in global shipping networks, in animal and plant bodies—can be rendered within its appropriating reach. Nicholas King characterizes the reordering of the globe around the project of securing emergent microbial risk as the "emerging diseases worldview."[38] A stock illustration, commonly used in numerous U.S. government publications on novel infectious diseases, suggests the scope of this refashioning. The image depicts a map of the world devoid of territorial borders or the names of nations, showing instead masses of land and water labeled by outbreak. Australia, for example, is labeled "Dengue, 1992." Mexico is "Hanta Virus, 1993," while Côte d'Ivoire is "Lassa Fever, 1912."[39] This image represents the world as a borderless orb in which emerging microbes move unfettered through the compressed networks of globalized realms. But that stock image and the mentality it casually emblematizes express more than fears over new patterns of contagion driven by globalization. The construction of the emerging microbes concept through the trope of a borderless globe should also be read in relation to a specific historical moment: the end of the Cold War, when the collapse of the Soviet Union left a relatively unarticulated and malleable world order in its aftermath.

A 2003 RAND publication, *The Global Threat of New and Reemerging Infectious Diseases: Reconciling US National Security and Public Health Policy,* reveals the role that the end of the Cold War played in the formation of the emerging microbes concept, while shedding light on late twentieth- and twenty-first-century American public health as a whole.[40] In setting the stage for its discussion of emerging microbial threats, the RAND document opens with an evocation of the end of the Cold War, described in existential terms: "The geopolitical landscape that now faces the global polity lacks the relative stability of the linear Cold War division between East and West. There is no large and obvious equivalent to the Soviet Union against which to balance the United States."[41] The report's neoconservative view of state identity was widely shared in right-wing circles at the time: there must be an external enemy in opposition to which a nation can be defined and its citizens unified.[42] In this sense, the Soviet bloc had been a reliable and enduring foe. If the Cold War was a conflict through which the United States had developed a sense of purpose and identity in the world, it also formed a large part of the basis on which the United States made itself socially, economically, ideologically, and materially viable for decades. The disappearance of the bloc from the world stage plunged America into a state of relative crisis, despite the apparent onset of victory and placidity. *The Global Threat of New and Reemerging Infectious Diseases* makes this scenario an occasion for a new kind of battle, a new sense of anxiety.[43] Although the decline of the Soviet bloc might seem to lead the world toward "the threshold of unprecedented peace," the report cautions that this concrete opponent is being "replaced by growing unease over non-traditional challenges."[44] Such national threats cannot be directly linked to the policies or actions of other states; without a foe like the Soviet Union, the world is "an environment in which it is no longer apparent what exactly can be done to whom and with what means." The report cites former CIA director James Woolsey's bleak description of this new terrain: "We have slain a large dragon, but are now finding ourselves living in a jungle with a bewildering number of poisonous snakes. And in many ways, the dragon was easier to keep track of."[45]

Since the world was no longer bifurcated by conflicts arising between two identifiable opponents, in a feat of global threat-casting, the RAND report represents the world as a space permeated with the unending risk of emerging microbes—a new class of foe. In contrast to the concreteness of the Soviet Union, this microbial enemy is "concealed" not by invisibility or covert forces, but by its very immateriality. RAND characterizes this new microbial rival as an "amorphous," "diffuse," "non-linear," and "grey-area" enemy arising in new and perpetually mutant forms. The report cautions against ignoring these nonconcrete risks by "subordinating them to prerogatives that are more concrete and more easily discerned," warning that doing so would render the nation incapable of meeting the challenges of the post–Cold War world.[46] In a now borderless world that is difficult to secure, the task is to counter emerging microbes before they can actually materialize.[47] In order to do so, RAND asserts that "the United States needs to revisit how it defines security and formulates mechanisms for its provision."[48] The report recommends that the United States focus on the indistinct and infinite menace of emerging microbes to redefine its purpose in the world. This new mission will also form the basis on which to maintain and expand America's international presence. Here, then, is an opportunity for the nation to give shape to an undefined post–Cold War order through a globalist project to fight microbial emergence.

Naming emerging microbes as the new enemy profoundly shifts the war logics of the American security state. This novel logic amounts to no less than a redefinition of war itself, reformulating the terms of conflict in at least five crucial ways. First, it changes what constitutes an offensive against the nation. In more traditional contexts, military interventions arise out of state-to-state conflicts that pose manifest threats. In the war on emerging microbes, in contrast, the danger need not be existent or probable, but only possible. Second, this transforms both war rationale and combat logics: the military strategies developed to counter microbial risk involve "getting ahead of the curve" of emergence,[49] acting against microbes in the present to preempt them before they can become an actual threat. Third, the emerging microbes concept alters the sense of the term *opponent.* In encouraging scientists to envision the "globe," the "world," the "biosphere," and the "planetary" as a space-time system pervaded with microbial potential, the discourse on emerging microbes redirects national public health efforts away from actual health threats within the state (such as diabetes, cancer, mental illness) toward biosecurity efforts to address perpetual pandemic risk. The national foe here is a formless force, a profusion of immaterial potentialities suffusing the world. Fourth, the U.S. war against microbial possibility cannot be fought on battlegrounds in delimited regions, but must encompass the entire globe and its systems. Fifth and finally, while war was once time limited—an event launched to achieve a desired outcome whose fulfillment would lead to a cessation of hostilities—this new war is endless. Battling all possible forms of microbial emergence means being suspended in a ceaseless security posture. At the same time, it means being locked into a drive to develop technological and scientific capacity in order to safeguard a future now perilously open to a hypothetical and immaterial opponent, one present nowhere but possible everywhere.

In order to counter a microbial emergence that suffuses all times, spaces, processes, and ontological categories, its threat must be first coaxed into resolution as an opposable and operable object with a visible, apprehensible, and calculable form. But the object of the war on emerging microbes is an evanescent target. That ambiguity works in favor of post–Cold War militarization, because the endless threat of emergence becomes a frontier by which the U.S. security apparatus can build itself without end. The project of microbial resolution must manifest an opposable object while maintaining its productively amorphous and unformed nature. This project reconfigures the planet as a new kind of problem-space whose threats must be anticipated through new capacities of visualization and mediation—materially, sensorially, technologically, and temporally—so that emerging microbes can be seen, foreseen, and fought.

But where to look for emerging microbes? And how to make the future appear in the present as a governable and securable object?[50] Through data visualizations of pandemic risk; computer modeled futures of possible pandemic outcomes; hand sanitizer dispensers; mobile phone apps that map individual and collective proximity to areas of reported illness; GPS-tracked animals to surveil zoonosis; genetics catalogues of viral matter; thermal sensors to detect fever stationed in airports worldwide; self-monitoring pamphlets; coughing and sneezing etiquette posters. I will return to some of these later, but here they offer just a sample of the new techniques and technologies, objects and practices that frame and focalize the extent to which vision and mediation are retrained around the speculative effort to glimpse and pin down microbial futures.

In keeping with this "world on alert" logic, such technologies and objects underscore the very impossibility of seeing emergence.[51] At the same time, they highlight the fact that amorphous "anywhere and everywhere" pathogens must be constantly surveilled. Yet the project of biosecurity operates in the gap between sight and foresight; the war against emerging microbes is a campaign against imagined futures. Because these techniques and technologies strive to mediate and visualize a temporally, spatially, and ontologically illimitable enemy, they generate projective, imaginative, and speculative ways of seeing and feeling. They work not by representing and scanning the concrete and evident, but by giving shape to the imagined and anticipated. In doing so, they endow the latter with the credibility of actionable information.

From Micro to Macro, from Biology to Ecologies

Let us consider five images that exemplify the optical operations of this emerging infections worldview (Figure 2.4). The first is a computer-modeled picture from MIT's "sneeze lab" that algorithmically calculates how infectious microparticles may travel across long distances. The second image, a diagram used at a National Institutes of Medicine (IOM) conference on air transport and emerging infections, maps out how infectious matter may be transmitted through air currents within airplane cabins. The third illustration, published in *Nature,* uses atmospheric modeling to determine how wind currents carry inhalable dust fractions of aerosolized poultry feces from infected birds from one farm to many others across state lines. The fourth image shows a map of

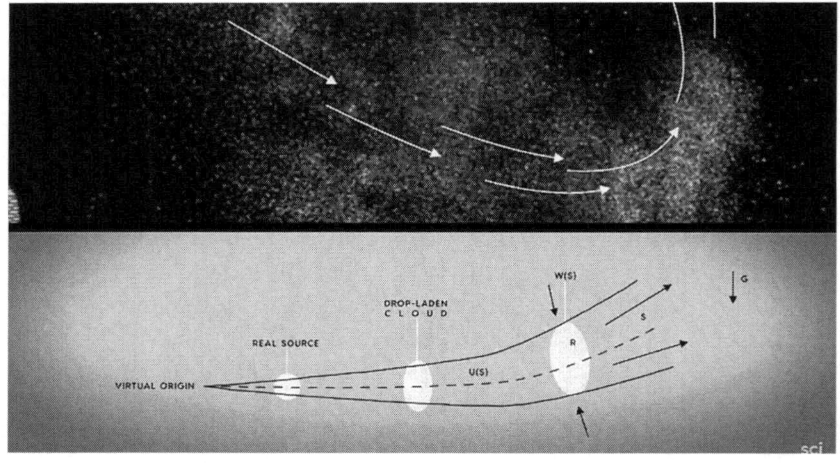

a

Figure 2.4. (a) Calculating and mapping trajectories of sneeze droplets using fluid dynamics. Screen capture from "Nothing to Sneeze At," SciFri Videos, 2014. (b) and (c) Studies of the movement of airborne droplets within the velocity fields of airplane cabins. National Airline Transportation Center of Excellence for Research in the Intermodal Transport Environment, "Infectious Disease Transmission in Airline Cabins," 2012; (d) Mapping wind flow to explore how air currents might have played a role in an avian flu outbreak in U.S. hatcheries. U.S. Department of Agriculture-National Institute for Food and Agriculture and Egg Industry Center, Iowa University, "Airborne Transmission May Have Played a Role in the Spread of 2015 Highly-Pathogenic Avian Influenza in the U.S.," 2019; (e) Mapping possible global networks of contagion by data visualizing global flightpaths, Digitiglobe/BioDiaspora, 2009.

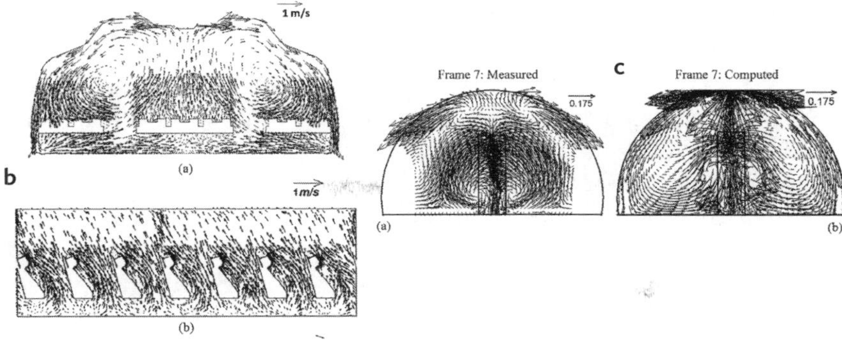
1 m/s
(a)
b
1 m/s
(b)
Frame 7: Measured
0.175
(a)
c
Frame 7: Computed
0.175
(b)

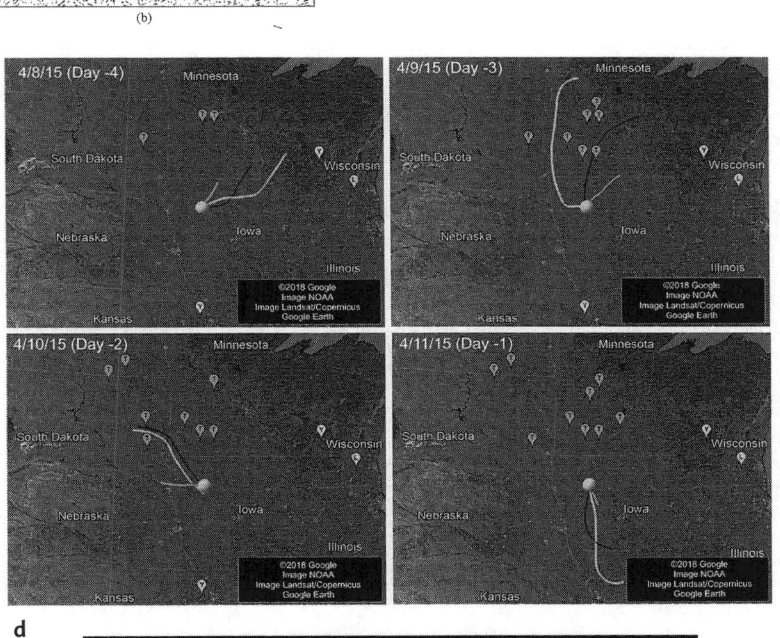
4/8/15 (Day -4)
Minnesota
South Dakota
Wisconsin
Nebraska
Iowa
Illinois
Kansas
©2018 Google
Image NOAA
Image Landsat/Copernicus
Google Earth
4/9/15 (Day -3)
Minnesota
South Dakota
Wisconsin
Nebraska
Iowa
Illinois
Kansas
©2018 Google
Image NOAA
Image Landsat/Copernicus
Google Earth
4/10/15 (Day -2)
Minnesota
South Dakota
Wisconsin
Nebraska
Iowa
Illinois
Kansas
©2018 Google
Image NOAA
Image Landsat/Copernicus
Google Earth
4/11/15 (Day -1)
Minnesota
South Dakota
Wisconsin
Nebraska
Iowa
Illinois
Kansas
©2018 Google
Image NOAA
Image Landsat/Copernicus
Google Earth
d

e

the world overlaid with a thicket of glowing lines to depict how far and fast contagion might spread via global air travel.[52] These studies are part of a broader research program that aims to anticipate global microbial threats by tracking aerosol particles suffusing the world. Considered together, the four images exemplify the ways in which we are incited to imagine potentially catastrophic microbes everywhere in our midst: dry coughs, wet sneezes, buoyant particles, aerosolized spittle, moist breaths, cold drafts, aspirated particles, turbulent air, atmospheres carried from one place in the world to another in the bodies of birds, animals, and flying metal cylinders. Air itself, the substance of atmospheric and pneumatic logics, becomes the medium through which the world is produced as a global space of perpetual risk.

The project of microbial resolution calls for the development of new capacities of perception to see the ambiguous threat of emerging microbes. These novel modes of apprehending the globe (and our existence in it) materially remake the world in the image of emerging microbes. The examples given also embody new analytic, imaginative, projective, and speculative ways of seeing. Technologies that explicate atmospheres operate in a constant interplay of fragility and insecurity. Even as they highlight air as a medium of our vulnerability, they also compel us to overcome that atmospheric anxiety by making it the site of technological interventions.[53] This explication of air "creates a vicious cycle: the more we learn about and exacerbate the vulnerability of the atmosphere, the more we turn to technology to compensate for and control the vulnerability."[54] Each of the four images registers this tension between the desire to control the atmosphere and the recognition that air resists any rendering as a predictable medium. At the same time, these images help to reframe the experiences of everything living in this once-familiar medium as fellow inhalers of its microparticulate registers.[55] The newly charged atmosphere is defamiliarized as an alien medium of disorientation, and the merely notional figure of the emerging microbe comes into sharper resolution as a material force suffusing the world.

In her analysis of ocean media, Melody Jue observes that when scuba divers sense they are losing control underwater, they attempt to "stand up," only to find themselves unmoored from familiar points of positionality.[56] Being suspended in the newly strange element of air enacts a positional relation parallel to the deep-sea diver's disorientation: a displacement that dislocates the body and defamiliarizes the senses, repositioning them within an alien context of incalculable threat.[57] In the era of emerging microbes, breathing itself—the most basic reflex of life—becomes as foreign, disorienting, and disconcerting as the magnitude of alien oceans or cosmological expanse. We are habitually unaware of the networks that constitute our daily worlds and experience. But the militarized project of microbial resolution reminds us that our everyday realms are what Anthony Giddens terms "phantasmagoric";[58] it compels us to see our daily domains as inhabited by "absent presences," "living but absent people, and present but dead or congealed processes."[59] This awareness foregrounds the fact that bodies are continuously enfolded into other bodies, scales, times, and systems through the automatic process of breathing. Once a familiar, unthinking, and life-sustaining element, microbial resolution recodes air as a scene of material displacement. The simple act of breathing implicates subjects in a world of microbial agglomeration and connection with global forces and entities,[60] the indecipherable ecologies of microbial transmission and emergence that Celia Lowe terms "viral clouds."[61] In this uncertain element, the wearing of facemasks, the suspension of breath, the swerve of the head when another person passes too close, are experienced as ways of being that materialize atmospheric menace. A once immaterial register drifts into resolution as an elemental medium that involuntarily draws all things on the planet together in mutating atmospheric enfoldings and unfoldings. The security project that seeks to detect and forestall emerging microbial life across the biosphere recodes both air and its respiration as dynamic materialities of potential transmission. And by cultivating anticipatory readings of the world through the molecular and microparticulate, the discourse on microbial emergence heightens this sense of ubiquitous risk, evoking a particulate seeing and intuiting of the vast range of deadly microbes that potentially circulate everywhere—literally "in the air."

In sensing the particulate, this discourse relocates and reconfigures "human encounters with the vital organismic agencies of bacteria, viruses, and fungi" on a planetary scale. Heather Paxson names human encounters with these assemblages "microbiopolitics."[62] I would also suggest the term *macrobiopolitics* to designate the expansion of biopolitical management to frame the planet and all it contains as a unified orb within a network of microbial commonality. Macrobiopolitics is not concerned only with microbes and our human encounters with them in the relational spaces of microbiopolitics. Instead, it refashions the biosphere itself as an inseparable realm of microbial flows in which organisms, ecologies, and objects are mutually implicated in the process of emergence. The emerging microbes concept forms part of this project to govern potential contagion in ways that depart from classical public health models; whereas those older regimes of public health focused on managing bodies, populations, and states, this biosecurity effort addresses the planet as a domain pervaded with unfolding microbes.[63] Indeed, this new approach crystallizes in the four preceding images, resituating the security problematic within the evolution of life across the biosphere. In this, it inhabits the molecular realm in order to reach the macropolitical in ways that broaden the meanings and places of war. Here, the project of microbial resolution continuously works to renew U.S. Cold War military values, infrastructures, and institutions in its endless pursuit of the spatiotemporally elusive frontiers of emergence.

The Transnatural Biosphere

Critical biosecurity studies often center on the ways that pandemic preparedness efforts seek to fortify the border between humans and other species. As part of this work, scholars have analyzed how biosecurity practices act on and use animal bodies as raced, classed, and gendered objects of waste and sacrifice: preemptively pumping farmed fish with antibiotics, burning heaps of animals deemed risky, fogging insects, gassing poultry, drowning civet cats, and more.[64] Yet the broader realm of microbial governance cannot be limited to a species-centered discourse, which tends to render peripheral the wider ecologies in which those species coexist. Looking beyond species can make legible the expansive footprint of U.S. militarization in the project of microbial resolution; indeed, the idea that a dynamic and interconnected world constitutes an uncontainable medium of emergence, and therefore also an expanding terrain to be secured, was central to the invention of the emerging microbes concept. In its landmark report on *Emerging Infections,* for instance, the IOM recommends that, in surveilling for deadly pathogens, biosecurity experts must keep a watchful eye over global space as a whole. This task cannot be accomplished by repeating the fragmented approach of the classic biopolitical paradigm, which focused on discrete populations in places such as those named by the IOM—Nigeria, Mexico, Japan, the Congo, or Papua New Guinea. In making its case for a truly global surveillance apparatus, the report stresses that, whatever the location, pathogens do not pose a threat on their own, nor do Nigerian sheep, Mexican hogs, Japanese fish, Congolese antelope, or New Guinean mosquitoes.[65] Instead, the real threat lies in the way that "human demographics and behavior, technology and industry, economic development and land use, international trade and commerce, microbial adaptation and change, breakdown of public health measures" have created "facilitative pathways"[66] over which microbes cross and mutate: an air ventilation network gives rise to outbreaks of listeria; golf courses encourage incidences of Lyme disease; discarded tires in a dump enable the rise of mosquito-borne encephalitis; war displaces people across the earth, leading to the resurgence of tuberculosis; travel enables flu microbes to reach across the globe in less than a day; expanding supply chains install new vulnerabilities in the global movement of goods and bodies; industrialized farming reshapes microbial life forms in myriad ways.[67] Rather than species, ecologies come into view as the wider arena of action under the project of microbial resolution.

By bringing this ecological background into the foreground, I highlight the ways that the war against emerging microbes prods biosecurity experts to see and treat pathogens not as present in discrete and bounded objects or bodies, not as narrow processes, but as inextricable parts of a complex, worldwide ecology. This larger system is composed of enmeshed human and nonhuman bodies, natural and manufactured systems, biological and technological processes—what I call the "transnatural biosphere."[68] If the

Pasteurians, according to Bruno Latour's analysis, saw microbes as enmeshed in all human and social relations while legitimating the hygienist's right to be everywhere within them, then the discourse on emerging microbes produces the transnatural biosphere as a new and widening realm that must be secured.[69] Tracing microbes through their ecologies, in turn, identifies the networks that must be surveilled, intervened in, and acted upon to forestall all possible future pandemics. This constitutes a dramatic deterritorialization of the biosecurity object, which mutates into a vast range of objects under a system of global monitoring: rubber boots on a hatchery in Idaho; migrating refugees at the Syrian border; supply chain routes crossing through Singapore; melting permafrost in Siberia; a meat processing plant in Mexico; an elevator button in Vancouver; bowling balls in New York; a transcontinental flight from Dubai.[70]

The shift from governing species life to securing the transnatural biosphere traces the new military-logistical landscape of this post–Cold War U.S. security project. Its operational arena no longer addresses an identifiable external enemy but is absorbed and diffused to identify one that is potentially internal and everywhere. Rather than engaging in combat in circumscribed spaces, U.S. post–Cold War warfare renders world ecologies geopolitically pliable by using the dangers in these systems to establish military and scientific conditioning and control.[71] For example, U.S. laboratories in the Pacific proving grounds served to consolidate U.S. military efforts through the invention of ecosystems science.[72] After the atomic bomb tests in the region, the Odum brothers did not go to the Pacific atolls to study the effects of radiation on ecosystems, as is commonly claimed; instead, they "traced radioactive isotopes in order to understand basic ecosystem processes, as if they were reading an X-ray."[73] Ecosystems were rendered visible as new objects of U.S. Cold War science and militarization by using radioisotopes as tracers.[74] The project of microbial resolution renews this strategy; resolving microbes as a perpetually emergent force that saturates the biosphere justifies the migration of warfare into earth systems themselves. Under this broadened sense of war, the endlessly risky frontiers of microbial interiors intertwine with the elusive frontiers of global emergence. As a result, U.S. post–

Cold War security imperatives take hold across a spatially and temporally unbound field of influence. By tuning into the flows of microbial possibility that flood the world, American-led combat extends through the globe and into its futures. At this point, it becomes legitimate to ask whether the project of microbial resolution mobilizes U.S. security apparatuses and systems in pursuit of a never-ending threat, or whether it reasserts U.S. military values, logics, and institutions precisely *because* there is no primary enemy in the post–Cold War period through which to renovate the U.S. security state.[75] As I elaborate in the coming chapters, either alternative renders that state capable of penetrating multiple registers of the world. What began as a quest to plumb the depths of microbial life on a boat in Alaska leads far out into a reconfigured biosphere.

As this chapter has shown, the concept of "emergence" works as a placeholder across an endless chain of possible futures, because emerging microbes are nowhere to be found in precise times or specific places. As the new targets of national biosecurity, RAND's "amorphous, diffuse, and grey-area threats" persist within the conditions of the world, and so the project of preemptive biopreparedness can spread through the world in their pursuit. This project produces warfare on a series of structural contradictions: it trains its militarized sights on an always-diffuse enemy; it is built upon the imperative that nonexistent enemies must be coaxed into resolution; its knowledge forms are structured by the material paradoxes of emergence; its security efforts produce insecurity; and it merges the trajectories of humanitarian global health with movements of global geopolitical strategy.

These contradictions form the lived backdrop of everyday life through an aesthetics of uncertainty. Under preemptive biopreparedness, these tensions and impossibilities infuse quotidian modes of being, feeling, and perceiving: the cultivation of the senses to feel out the tensions of the world through the microparticulate, the (mis)attunement to once-familiar air as an element now suffused with an alien charge, the inescapability of a crisis that is endless, everywhere, and always possible. We can register some of these effects in the mundane "just-in-case" affects of the airplane traveler, and the "just-in-case" logics of others (like me) in the previous chapter. Through the aesthetics of uncertainty, subjects rehearse the gestures of preemptive biopreparedness, live in its crisis ordinariness, and train our perceptive horizons and capacities toward the catastrophic future pandemic. This quotidian practice immerses us in conditions of radical contingency; facing an everywhere and always possible enemy, one experiences the failure of vision to apprehend its subject, the inability of scientific instruments to stabilize risk, and the incapacity of expert systems to hit their mark. As a result, we must reckon with a breach of confidence in the systems, techniques, and technologies in which we once placed our trust. Part 2 of this book will explore some of the crucial ways in which scientific knowledge production has been remade in the moment of its insufficiency.

Part II

The Calculative Imaginary

3 Flightlines and Sightlines

The wild birds inhabiting the regions around Mainland China's Lake Qinghai and Poyang Lake seem ordinary enough; they fly, feed, migrate, mate, and nest like others of their kind. But several species like the bar-headed goose and the wild ruddy sheldrake have GPS devices harnessed to their backs (Figure 3.1). Wild migratory birds have become key objects for researchers seeking to understand how an avian flu virus could evolve to wreak havoc on a global scale. Many scientists believe that avian flu viruses become more virulent as they pass between wild and domestic birds, posing a significant threat to human health and survival. Drawing on arguments made in the 1960s and 1970s by scientists such as Robert Webster and Kenney Shortridge, who suggested that China could become an epicenter for future influenza viruses, later researchers designated the regions around Lake Qinghai and Poyang Lake as areas of concern due to the proximity of over a million wild migratory birds to numerous backyard farms and wet markets.[1] The research groups using these tagged birds include both international and Chinese scientists who work across the fields of wetland conservation, veterinary medicine, ecosystems science, economics, and climate and spatial modeling. Their interdisciplinarity reflects the terrain with which these scientists are concerned: they are not studying the viral phylogeny or ecologies of avian flu (approaches that examine the genetic lineage of viruses), but rather its disease networks (that is, the networks of relations and conditions through which viruses evolve).[2]

This project began because researchers wanted to understand what factors made China so ecologically hospitable to avian flu

Figure 3.1. A bird with a GPS tracking device harnessed to its body. U.S. Geological Survey, 2016, photograph by Michael Casazza, Western Ecological Research Center.

outbreaks. Using these bird technologies, researchers hoped to test the widespread and broadly accepted hypothesis that presumes China as the ecological birthplace and epicenter of a future highly pathogenic, human-transmissible avian influenza pandemic.[3] Scientists hoped that by tracking these birds they could pinpoint where, when, and how wild birds interface with domestic animals and humans, and it was thought that wet markets would be key sites through which to track such crucial cross species overlaps.

In keeping with the anxiety over the atmo-political dimensions of emerging microbes and their capacity to blur species, scales, and temporalities, GPS tracking is used to trace the ecological enmeshment of these wild birds within larger transnatural systems. GPS devices allow researchers to track the migratory routes and stopping grounds of tagged birds by translating their movements into digital images that reveal their flyways (Figure 3.2a). The resulting data is layered and compiled with scores of other data banks that show researchers what's on the ground below these migratory paths: land surveys of terrain and hydration periods,

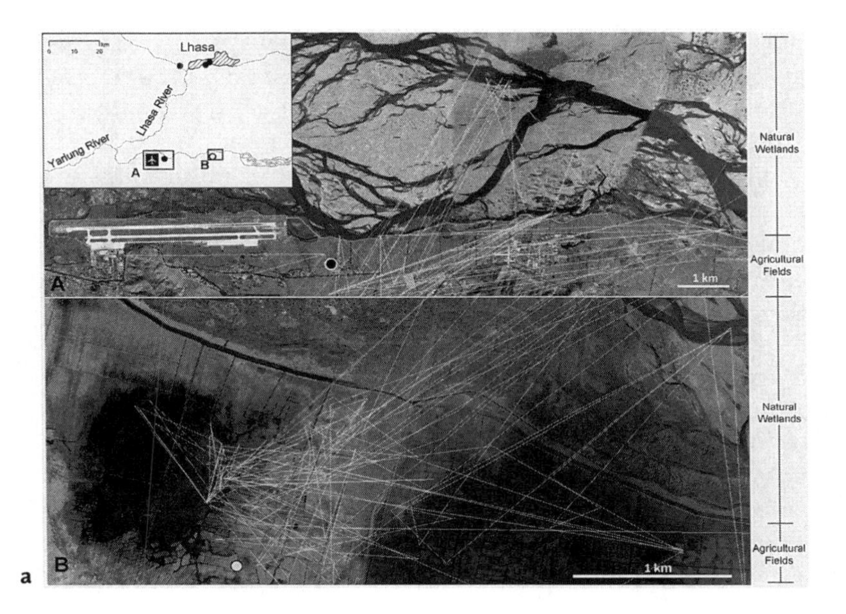

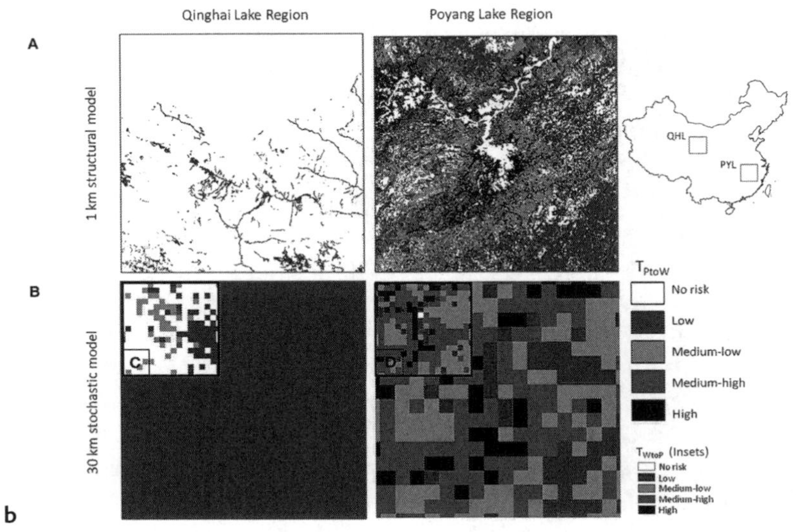

Figure 3.2. (a) Tracking bird flyways in the Qinghai-Tibetan Plateau using GPS-harnessed birds. U.S. Geological Survey, Patuxent Wildlife Center, Diann Prosser et al., 2011; (b) Data visualized maps displaying calculated levels of pandemic risk in regions around Poyang and Qinghai Lakes. U.S. Geological Survey, Patuxent Wildlife Center, Diann Prosser et al., 2013.

maps documenting the location of human dwellings, rice fields, and domestic poultry farms. Finally, all this data is used to measure levels of pandemic risk by superimposing gradations of colors over a map of a specific region (Figure 3.2b).

In the previous chapter, I showed how emerging microbes formed the basis of a new kind of post–Cold War security object that suffuses the world with ever present risk because it lacks the clear-cut and concrete nature of the Cold War opponent. Here I turn to the use of these birds and the maps they generate to examine how the diffuse and uncertain nature of emergent microbes is coaxed into visibility and transformed into a governable object of security. I analyze these tagged birds as an example of risk technologies—technologies that generate information out of the unknowable to enable action in the face of radical indeterminacy—and explore the epistemological forms called up by the project of preemptive microbial governance, in which *uncertainty itself* takes a central place in contemporary knowledge systems. Many components make up these GPS/bird/environment/biotechnological assemblages; I group these as a single technology that I call "animal sentinel media." Anthropologists such as Frédéric Keck and Lyle Fearnley have examined the use of wired birds in their studies of contemporary biosecurity.[4] However, in framing these avian bodies and their systems *as media technologies*, I set them within a longer history of devices constructed to bring the uncertain and unseen into the realm of knowledge, and position them in a larger set of questions to probe the interactions of vision, science, and their instruments. I call them "sentinels" because one definition of the term is "the soldier who advances ahead of their troops in a battlefield to see what dangers lie ahead."

Animal sentinel media raise questions about the relationship between data and emergence: How are mutable futures brought into being through data? This chapter follows the concept of resolution as an optical process that clarifies hazy details, and converts abstractions into graspable forms to probe these questions. I consider animal sentinel media as they visualize uncertainty, and through this function, I track a shift in the making of indexicality in twenty-first-century epistemologies of risk. If the familiar formulation of scientific imaging links vision to knowledge (that is, if the ability to see leads to the ability to know by illuminating, uncovering, or representing), I examine how the project of microbial resolution converts emergence into a security object by forging a novel relation between sight and foresight. I pursue this argument in three steps. First, I situate animal sentinel media within a longer history of imaging technologies that seek to bring the invisible phenomenal world into visible knowability by establishing an indexical relationship with that world. In drawing out some of the continuities and discontinuities between these technologies and animal sentinel media, I foreground the ways in which the speculations generated from such technologies achieve their status as scientific knowledge. Second, I focus on the risk maps produced through such animal sentinel media to explore what it means to make a picture of risk itself: to construct an index of a time, a place, and an event existing only as pure possibility. Finally, I explore animal sentinel media in their function as risk technologies whose very role is to produce knowledge of the anticipation of unknowable futures. If risk is unknowable, immeasurable, and incalculable, animal sentinel media are devised to render it subject to quantification and measurement. I analyze the anticipatory drive underwriting these animal sentinel media (a drive I link with the speculative logic of the medium as a data mining and processing technology), and ask what it means to make emergent events into visible objects of knowledge when that process can only be replete with failures in the form of error, noise, incompletion, and nonknowledge. Yet within twenty-first-century knowledge systems, failure and error have themselves become "qualities" to be incorporated and embraced rather than eliminated.

Although it is tempting to implicate such technologies in the creation of "geographies of blame," through which racialized people in "distant" lands with "exotic" customs and "uncivilized" cuisines are blamed for the existence of disease,[5] I defer this analysis to the following two chapters. As Fearnley shows in his extensive and illuminating study, as research groups tried to pinpoint the epicenter of a future avian flu pandemic at the interface of wild and domestic birds in China, these attempts actually led them in the opposite direction. Among other things, through their research, they discovered that the categorical divisions between "wild" and "domestic" became blurry; that the proposed interfaces of viral spillover did not hold; and that intensified industrial farming is most likely what renders geographies susceptible to the generation of avian flu viruses.[6] Rather than confirming the "China hypothesis," then, these experiments eventually displace it. And while there are indeed racialized and neocolonial logics in the management of contemporary pandemic risk, I hope to open up more nuanced dimensions in the remaining two chapters. First, however, this chapter will draw attention to the ways in which the complexities of data-driven sight have changed the very definition of what it means to know something.

This chapter traces the technological reconfigurations of vision in relation to questions of knowledge. These technologies aim to locate the possible origins of future avian flu pandemics in order to intervene in them in the present. In this way, and whatever other work they do, they are also signature technologies of preemptive biopreparedness. The ways in which GPS-tagged birds produce meaning and generate information call for analysis within this broader framework. Within this framework, these birds are used to fathom and explore microbial life, leading researchers to lay out the future as a proliferation of possibilities—an ever-expanding imaginative horizon of what is possible, even if not probable. If probability uses statistical calculation to determine what is most likely, this chapter explores how animal sentinel media work beyond probability, weaving uncertainty into knowledge in a way that privileges possibility over probability. One key goal and effect of preemptive biopreparedness is to render that space of possibility operable. I show how animal sentinel media form a picture of emergence inscribed by and composed of data, quantifications, and measurements, indexing risk to make the emergent qualities of microbes into a workable security object. As Orit Halpern argues, "it is precisely this merger between [computational] vision and the reformulation of reason that underpins contemporary biopolitics."[7] This chapter then considers how the project to secure emergent microbes takes shape through data-driven technologies in ways that reconfigure vision, rationality, and the relation between the two. I show how this breakdown of vision itself arises as a novel mode of observing the unknowable and, through these traits, I trace the rise of the "calculative imaginary": a proliferation of numeracy, informatics, quantification, and measurement precisely where vision reveals its inadequacy. The calculative imaginary emerges as a rhetorical, epistemological, and aesthetic convention to address the incomprehensible magnitude of catastrophic risk in the twenty-first century. I examine the extent to which uncertainty forms the basis of a new rationality and reflect on the epistemological and material implications of scientific knowledge systems that take shape around the uncertainties of contemporary cultures of emergent risk.

Sight and Foresight

The quest to fix, with highest fidelity, the fugitive and imperceptible phenomenal world into indexical traces animates the long history of scientific visualization. Étienne-Jules Marey tracked on camera and calibrated in measurements the incremental undulations of birdwing movements to understand the mechanics of natural flight. Eadweard Muybridge, in turn, famously trained his photographic camera on a running horse to settle a wager on whether horses maintain at least one foot on the ground when galloping. Hermann Schnauss captured invisible electric fields in electrographic images of sparks emanating from wires and bodies (Figure 3.3). The telescope, the microscope, the cinematic and photographic cameras, and related technologies have all been used in scientific imaging as tools to assist in the production of knowledge.[8] Their presumed neutrality is the condition on which the "truth claims" of such images are made.[9] We might call these devices "technologies of certainty": by transforming the imperceptible world into indexical images, they claim to provide an account of something true, despite numerous discussions around media and vision that rightly contest the reliability of the index.[10] However, when it comes to the making of scientific knowledge, the index is still central to its imaging practices and remains a large part of how proof is constructed.

Scientific ways of seeing seem to offer reassuringly authoritative knowledge by drawing the unseen into visibility. Kirsten Ostherr argues that films about global contagion operate according to a

Figure 3.3. (a) *birds,* chronophotograph, Étienne-Jules Marey, 1882; (b) *Jockey on a Galloping Horse,* Eadweard Muybridge, 1887, National Gallery of Art, gift of Mary and Dan Solomon and Patrons' Permanent Fund; (c) *Electrographie eine Drahtlehre aus Messing,* Hermann Schnauss, Albertina/Höher-Graphisches Bundes-Lehr-Und-Versuchsanstalt.

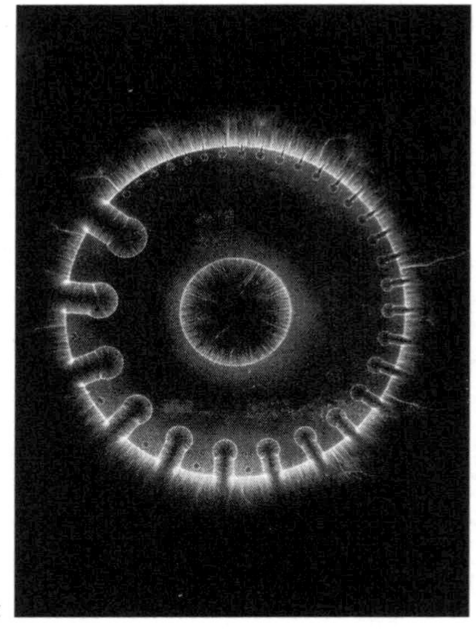

similar logic: "If one can see the contaminant, one can avoid the infection." In this formulation—what Ostherr calls "representational inoculation"—seeing is equated with knowing and preventing.[11] Yet the designation of viruses as *emerging* means that the coming pandemic exists in a state of suspended nonknowledge. Here we meet the limit of the classic formulation of scientific visualization that equates sight with knowledge, for how do we see *potentialities*? Or, as one epidemiologist formulates it, how do we build a "system that can detect the novelty" of pandemics that have yet to manifest? The project of securing microbial emergence faces a visual problematic unaccounted for in Ostherr's formulation. Departing from this framework, I ask instead: What is the relationship between sight and foresight?

This question involves a puzzle of how to relate the visible to the uncertain: how can something be indexed when the referent for which it purports to stand in does not exist? Animal sentinel media, used as data visualization technologies, make images not from the transcription of the real world, but from bits made of binary code. Animal sentinel media are believed capable of making a picture of risk through data itself. The images that these birds make begin, then, with a material abstraction that forms the essence of the digital, as Aden Evans puts it.[12] Have we arrived at what many writers on digital media (such as David Rodowick, Lev Manovitch, and Jean Baudrillard) describe as the death of the index in the digital era?[13] In these discussions, digital images no longer adhere to the tradition of technologies of certainty because they manifest an assembly of code rather than an index. In such cases, digital imaging is often conceptually untethered from the traditions of photography and aligned with the history of painting.[14] However, the index has not in fact died with the digital. Instead, it has been relocated.

Rather than transferring the optical conditions of luminosity onto a light-sensitive surface, as with the photographic or filmic index, the bodies of the birds themselves become the index as animal sentinel media. Unlike the laboratory test animal whose flesh must be cut off from any relational ties to its surroundings (so that the body can be purified and act as proxy for some other entity), the sentinel animal's functions rely on its living relationality and embeddedness within larger planetary systems.[15] This functional difference is undergirded by a notion that, by virtue of the very material entanglement of the sentinel body with broader processes and phenomena, it can lay bare its specific milieu of existence. Like the canary in the coal mine that detects the presence of poisonous gases (and whose body serves an acceptable sacrifice in the place of a potential human one), sentinel devices put birds' bodies to work as feeling, perceiving, and sensing mechanisms. Indeed, one proposed etymology of "sentinel" is "to feel and perceive through the senses." What they are made to feel, sense, and pull into human perception is the literal tracing of their flyways as the contagious global network through which a future pandemic might emerge.

The Promise of Pure Perception

According to this sentinel logic, avian bodies are technologies of pure perception, and the data visualizations produced from their indexing bodies appear to offer an image of the future free from the complicating, willful interference of human rendering and processing. These visualizations are seen as capable of accounting for otherwise ungraspable, unknown registers. Their collective, corporeal immersion in the planetary networks that enable pandemic possibility allows them to operate as perceptive instruments capable of indexing these networks with the highest precision. Avian bodies are taken to be reliable repositories and transmitters of information because of the "natural" intelligence thought to inhere in their immersion in larger networked ecologies and systems.[16] This immersion ostensibly promises a zero-space/zero-lag between data and its image, which endows the resulting data visualizations with a sense of communicative objectivity and indexical credibility. What was once the perceiving lens of the camera and the photographic plate converges here within the bodies of birds. Accordingly, these visualizations are supposedly able to obtain the documentary imperative of impartiality, surpassing previous efforts muddied by the accidents and agendas of individuals and institutions. Such imaging seems to loosen at last what André Bazin calls the "sin" of subjectivity, the crucial shift in the place and presence of this index.[17] As researchers track these birds' movements, they use these bodies as listening posts into emergent futures. In using their bodies to fly first into the viral future ("going first" is yet another definition of "sentinel"), animal sentinel media move through their pathways while their indexing flesh is deployed to sense the edges of the pandemic horizon. Their flyways are traced in bright glowing yellow and orange lines, and researchers use this information to disclose the networks that connect the Google Earth images with other regional data points. By analyzing and correlating such data points to make cartographies of pandemic risk, these flightlines become indexical sightlines that guide the scientific gaze to a view of mutable pandemic futures. Rather than pointing to a specific moment, thing, or probable scenario (the index that points "here," "this," "then"), or to a place (the map that delineates a site), the risk maps generated by these avian bodies point instead to an "elsewhere," to the "not-yet-here" proliferation of futures yet to come. But in what form do the nonknowable and unfixed futures of pandemic possibility come to appear?[18]

The future, it seems, appears in the form of "raw data." This widely used term implies that information merely exists in the world, that it's simply "out there" like sand or coal, an endless resource awaiting discovery, extraction, and use. Such cultural fantasies about data as another kind of natural resource are clearly articulated in IBM's 2010 commercial, "Data Anthem."[19] The video opens with a scene of thousands of luminous dots floating against the black backdrop of an urban night sky as a voiceover declares with wonder: "Our planet is alive with data" (Figure 3.4). The viewer is then led into an animated world as a mobile, aerial observer, flying though a space that glitters with beautiful, abundant, and endless flows of data emanating from buildings and babies, electrical grids, and highways. All the while, a narrator reels off some of the innumerable sources of this plentitude. Data, the commercial suggests, is everywhere, and it has always been present. Like some secret and hidden language murmuring through the world, it has been waiting to be heard. Until recently, we lacked the right tools to listen to it. An extension of the techno-colonialist view of the world as a resource to be apprehended by technology, "Data Anthem" articulates the fantasy that the planet's seemingly infinite,

Figure 3.4. "Data Anthem," IBM Documentaries, 2010.

naturally occurring raw data can be transformed into something usable because we now have the right technologies. As the unseen narrator intones, this is how we "make data work for us."

The vision of a world alive with raw data to be put to productive use is powerfully at work in animal sentinel media. The sentinel body labors as a kind of corporeal resource, a data-driven technology that tracks points and makes correlations, enabling and contributing to the ceaseless, frictionless circulation of information in a world overflowing with data. Like the electronically tagged and tracked bodies of logistics workers (such as Amazon warehouse employees), their traced patterns, rhythms, and velocities of displacement enclose their flesh and its movements in the managerial and surveillance captures of informatics capital. And all the while, these avian bodies both generate and harness wellsprings of data,[20] delivering ecologies and their futures to the highly securitized and surveilled realm of microbial governance. Yet as Lisa Gitelman and others argue persuasively, "raw data" does not simply "exist." Instead, raw data is an informational concept manufactured by humans. The data generated through animal sentinel media must not only be conceptualized, it must then be processed, smoothed of "noise," weighted, categorized, and shaped by inquiry.[21] The belief that the planet is bursting with raw data opens it to technological development worldwide in ways that retrace old colonial histories of territorial conquest and resource extraction, making unsustainable material demands on the planet. This is a topic I examine in the following two chapters; to get there, however, I want first to point to the logic by which animal sentinel media are thought capable of conjuring up incipient futures from a world alive with data.

Cartographic scientists working in the World War II era were deeply influenced by the information paradigm shaped by cyberneticians' belief that the world was made up of proliferating data. In his account of informatics and mapping, Antoine Bousquet shows how, during that time, cartographers began to think of maps as communication channels that connect a map, its maker, and its user.[22] I would add that the belief that the world could be reduced to informatic relays, and that geoinformatics mapping technologies could connect humans to them, is rooted in that same historical moment. The risk maps generated through animal sentinel media develop out of this history; they are not only measurements and representations of space, but they also function as communications conduits, "portals to a dynamic, interconnected, and distributed world of data" referenced to sites and networks.[23] The indexing bodies of animal sentinels are used to tap into and visualize a world made up of ever-changing data flows. In this context, animal devices—and the maps that their bodies are used to make—transform raw data into value, and render them into useful data points that work for the mythical global "us," as in IBM's anthem. From this mythology a world teeming with naturally occurring streams of data, animal sentinel media are deployed as instruments that "pick up" information from the broader environments in which they are embedded. This data derives its objective status as pure index through the natural intelligence thought to inhere in these avian bodies.

The Incipient Edge of Risk

Natural intelligence harnessed, perception unmuddied by human cognition, lines traversing real space, trajectories tracked in real time, nodes patterned, elements pinpointed, distances measured, scales counted—for all this accounting, the true subject of risk maps is risk itself: the act of anticipating events that have not yet come to pass, and are therefore (per Beck) "immeasurable," "incalculable," and "unforeseeable." In this sense, animal sentinel media are tools that render uncertainty apprehensible through the transcription of the data they both generate and collect. These avian indexes work by what Charles Sanders Peirce calls *abductive reasoning*—a mode of reasoning rooted in premises and outcomes that may or may not materialize and enable the elaboration of a claim about uncertain futures. Whereas induction infers larger conclusions from specific cases and deduction draws conclusions from known principles, abduction involves provisional hypotheses that attempt to "fill in" gaps in knowledge in the face of unexpected, new, and unfolding findings.[24] The use of animal sentinel media resolves this gap into a regional map that measures levels of future pandemic possibility, assigns risk calculations, and expresses risk values in gradations of blue and green. Through their bodily script, the uncertain futures of possible pandemics are assigned a perceptible shape: plotted across distances, quantified and calculated, transformed into data points, and visualized as places, networks, and processes charged with risks to be governed in the present. As Tess Takahashi states, such data-driven images derive their documentary power from the data's capacity to stand in for an omniscient and omnipresent viewer who can account for everything.[25] The close relationship between the seen and/as data that endows risks maps with evidentiary force. Uncertainty is rendered into an image as "risk," and thereby takes shape as a more operable object.

What is the character of the operable object? And how are its traits structured by its indeterminacy? If the traditional index posits "here" and "this has been," the index of risk directs an observer's gaze to something that has not yet come to pass. The thing that the index of risk makes visible is structured by the impossibility of marking its object. Here, vision is structured, not by the optical properties of microbes themselves, but instead, by the unforseeability of their emergent futures. The index of risk generates sight from this fundamental condition of blindness. While the project of microbial resolution works to transmute the abstractions of emergence into visible and graspable form, these modes of visualization underscore the paradox inherent in this effort. Represented in and as points of information, data bites, measurements, numerical lists, and unresolved pixels *as markers of things not seen,*[26] the risk map presents a form of countervisual observation in which perception is pieced together as bits (*solver*: dissolves, breaks apart, disperses). This contradictory visual trope of sight as simultaneously epistemological claim and as perceptual blockage mirrors necessary incongruities in statements like those made by Anthony Fauci, director of the National Institute of Allergy and Infectious Diseases at the National Institutes of Health:

> Despite extraordinary progress during the past two decades . . . the perpetual nature of the emergence of infectious diseases poses a continuing challenge, which is volatile and ever-changing. . . . The future is ever uncertain, because unimagined new diseases surely lie in wait, ready to emerge unexpectedly.[27]

What can we make of the tension between the deluge of information provided by animal sentinel media, and the fact that we remain uncertain as to how, when, or where the next catastrophic influenza pandemic will occur? The flood of information constitutes the rise of what I call "the calculative imaginary"—the proliferation of numeracy, quantification, data, charts, maps, and informatics as a reconstitution of sight where vision meets its limit. The calculative imaginary surfaces in these risk maps as a mode of exploration that proceeds through optical inaccess. Vision (almost always also an epistemological conceit), as it is reformatted in the calculative imaginary, enfolds indeterminacy and uncertainty as modes of sight, and as a surfacing epistemological contention. Yet, *evidence*—what might be characterized as "fanatical descriptive clarity" or a "hallucinatory wealth of detail"[28]—is what the unresolved risk map refuses to reveal despite the promise of data-driven technologies to see everything from all points. Animal sentinel media are used to make these maps of future threat, but what does it mean, then, to index risk?

Risk maps are images of the dissolution of vision, of perception standing precariously at the boundary between the known and the nonknowable. What can be seen at the threshold of the unresolved and the uncertain? This problem of vision recalls the

Figure 3.5. (a) Trevor Paglen, *Detachment 3, Air Force Test Center, Groom Lake, NV; Distance approx. 26 miles,* 2008. C-print, 40 x 50 in. Copyright Trevor Paglen. Courtesy of the artist, Altman Siegel, San Francisco and Pace Gallery; (b) Trevor Paglen, *Open Hangar, Cactus Flats, NV, Distance ~ 18 miles; 10:04 a.m.,* 2007. C-print, 30 x 36 in. Copyright Trevor Paglen. Courtesy of the artist, Altman Siegel, San Francisco and Pace Gallery.

Limit Telephotography series of contemporary artist Trevor Paglen. In this body of work, Paglen stations his camera, outfitted with a powerful telephoto lens, at the outer borders of secret government sites. These spaces are "black sites," absolutely hidden from public view. When searched on Google Maps, these sites are thoroughly blurred out or covered over in tan blotches. When approached by foot, they remain invisible, surrounded and encased by miles of secured buffer zones. Fixing his lens on these black sites for long exposures across extended distances, Paglen produces images that register whatever traces his camera picks up. To cite two examples from the series, *Detachment 3, Air Force Test Center, Groom Lake, NV; Distance approx. 26 miles* and *Open Hangar, Cactus Flats, NV, Distance ~ 18 miles; 10:04 a.m.,* show an unfocused nocturnal landscape with a thin band of hazy light (*Detachment 3*), or pixelated images of unidentifiable objects (*Open Hangar*) (Figure 3.5). In Paglen's telephotographs, vision operates to lay bare its own limit. These are pictures of sight breaking down at the boundaries of the visible; we know that we are observing a mystery, but we can see neither its nature nor its content.

Similarly, the pandemic risk map expresses the border of "potential risk." Its zoomed-in detail—appearing in the lower register of the image as pixels of color excerpted from the larger map—is an image of optical straining, of looking too hard, to no avail and to no end. The term *resolution* can refer to the technical process of coaxing an optical detail into fuller view, to the process by which something seen poorly at a distance becomes clearer up close, or to the degree of detail visible in a mechanical image. The project of microbial resolution works in the same senses, but to different effects. As animal sentinel media are used to render emergent microbial futures, they produce images of vision breaking down. The risk maps drawn from their data visualize the future in an image of obstinate nonresolution. The computational vision that underwrites these technologies sees so intensely that sight falls apart into pixelated terrains. Zooming in to get a closer view of risk reveals an image of greater irresolution. Here, in the image of a low-information dpi map, is an image of a future.

Indexing risk and "better" capturing its phenomenal trace is not a matter of sharpening technique or of improving optics. These are maps of *risk itself.* They embody the ambiguity and difficulty of the quest to pin down and ultimately locate unknowable viral futures. They are the visual equivalents of statements made by specialists like biosecurity expert Michael Osterholm, who, in contemplating the avian flu virus, remarks: "Mother nature is trying to tell us something here. But I don't think any of us have a clue."[29] As such, despite all the information it gathers, taps, and makes visible, the risk map is similar to Eugenie Brinkema's characterization of point-of-view. For Brinkema, point-of-view is not only a technique used to convey a subject's perspective; it is sometimes used to stage before its proxy viewers the emptiness of a scene, organizing an image of what is not there. In such cases, the image reveals "the failure of visibility," serving not to provide an image of psychological or subjective truth, but showing "blocked vision."[30] Similarly, these risk maps—in their metaphorical and visual flatness and opacity, perhaps even in their visual boringness, for some—stage risk as a presentation of the inability to see. It is not so much a matter of seeing the unknown as of seeing that we can neither see nor know. The microbial future is presented not in the calculable, quantifiable, and delimited presence of the data from which its risk thresholds are determined, but instead in the map's presentation that the potential threat is *not* rendered present in the referential capture.

If animal sentinel bodies can index risk as the perpetual shuffling of potential futures, the risk maps derived from their bodies stage that uncertainty. These mappings of risk itself are the visual presentations of unknowable hypothetical disasters being played out, presented, and plotted in real space.[31] Their cartographies arrange the substanceless nothings of impossible-to-know futures as a tableau across space and time, rendering a picture of inevitable, but constantly deferred, future catastrophe. The hues of risky greens and blue compose an image of time as unending duration in that they point only to the minimal distinction between what may or may not be possible.[32] If the index is a finger that points "here," the index of "risk" punctures those flat opacities of color; rather than sustaining the barest separation between the present and future unknowns, they give way to bottomless and ever-changing microbial futures.[33] The indexing of risk refutes the conceits that the cartographic enables geospatial measurement and control, and that data can account for everything. In a striking departure from standard cartographic technique, the map is not an image of what has been, what is, or even what will be. Indexing risk is instead a visual testament to an open-ended emptiness, an unyielding depth filled with an unyielding uncertainty. In making a picture of risk in these maps, we see an image of human biological vulnerability, without clearly defined contours. That ambiguity is a generative space. The rendering of future uncertainty in this map enables the war against microbial emergence to refashion present transnatural networks into a future territory to be secured. Once risk is indexed, the gaps between a present condition and a future time can be filled in by envisioning catastrophic biological events in possible futures. Indexing risk entails making possibility itself appear in an image of what is not there. Indexing risk itself, in this sense, is an exercise in strategic imagination.

Imagination, the Operative

"Imagination." This word heads the initial section in the chapter on foresight in the *9/11 Commission Report* released in 2004 (Figure 3.6). In that document, beneath the title "Foresight—and Hindsight," the authors argue that the surprise attacks of 9/11 "reveal a failure in our faculties to exercise and deploy our imagination."[34] Hindsight, they contend, supplies us merely with the bitter taste of 20/20 vision after the fact. For this reason, the authors recommend that we "routinize, bureaucratize and institutionalize imagination" to avert other future security shocks. Only by fanning the embers of the imagination can "foresight" be achieved. Consequently, strategies to nurture and prod the national imagination and enlist it in a project of everyday militant vigilance assumed a central place in the preemptive phase of "the war on terror." Such a strategy was largely cast as an optical operation. This was the time of "if you see something, say something," a doctrine that encouraged its subjects to imagine impending danger in mundane objects: an unattended backpack, a locker at the bus terminal, a forgotten parcel. It was also the time of the U.S. Department of Homeland Security terror advisory color chart, which measured the risk of future catastrophe in five different hues (Figure 3.7). Masco asks of this terror advisory chart: "How is a 'guarded' threat to be distinguished, technically, from an 'elevated' one?"[35] We might pose an analogous question of the risk map: How is "high risk" to be distinguished from "medium-high risk," or "medium-high" from "medium-low"? In each of these examples, imagination intervenes in the blank space of the ambiguous to change the way we look at nothing in particular. As Henrik Gustafsson, Marieke de Goede, and others argue, by enlisting imagination in the always-surfacing conflict of preemptive war, the 9/11 report supports a logic in which the imagined precedes reality, which, in turn, is reshaped.[36] While the index must stand as evidence of something real, actual, and manifest, preemption instead organizes a prepositional world

11
FORESIGHT—AND HINDSIGHT

IN COMPOSING THIS NARRATIVE, we have tried to remember that we write with the benefit and the handicap of hindsight. Hindsight can sometimes see the past clearly—with 20/20 vision. But the path of what happened is so brightly lit that it places everything else more deeply into shadow. Commenting on Pearl Harbor, Roberta Wohlstetter found it "much easier *after* the event to sort the relevant from the irrelevant signals. After the event, of course, a signal is always crystal clear; we can now see what disaster it was signaling since the disaster has occurred. But before the event it is obscure and pregnant with conflicting meanings."[1]

As time passes, more documents become available, and the bare facts of what happened become still clearer. Yet the picture of *how* those things happened becomes harder to reimagine, as that past world, with its preoccupations and uncertainty, recedes and the remaining memories of it become colored by what happened and what was written about it later. With that caution in mind, we asked ourselves, before we judged others, whether the insights that seem apparent now would really have been meaningful at the time, given the limits of what people then could reasonably have known or done.

We believe the 9/11 attacks revealed four kinds of failures: in imagination, policy, capabilities, and management.

11.1 IMAGINATION

Historical Perspective

The 9/11 attack was an event of surpassing disproportion. America had suffered surprise attacks before—Pearl Harbor is one well-known case, the 1950 Chinese attack in Korea another. But these were attacks by major powers.

While by no means as threatening as Japan's act of war, the 9/11 attack was in some ways more devastating. It was carried out by a tiny group of people,

339

Figure 3.6. First page of chapter 11 of the *9/11 Commission Report.* National Commission of Terrorist Attacks on the U.S., 2002.

a

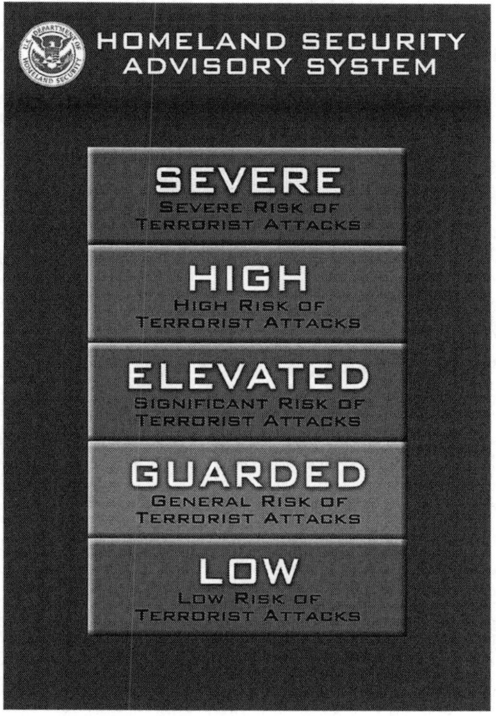

b

Figure 3.7. (a) National Terrorist Advisory System poster, U.S. Department of Homeland Security; (b) Homeland Security Advisory chart, U.S. Department of Homeland Security.

in which the index may be imagined as an unresolved projection. Here, paradoxically, the imagined may take on a more concrete existence as the basis for anticipatory interventions in the present. The project of preemptive biopreparedness emerges, in large part, from this history. Managing microbial emergence and preempting terror share the same project of strategic imagination. Both make foresight, fashioning it out of the capacities of a militarized imagination on the offense in a war game with high stakes. As in the terror chart, the colored patches on these maps are visual shorthands for proximity to catastrophic risk.

These cartographies of microbial risk transform the unknowable and imagined into an operable and workable borderland. The U.S. military calls spaces characterized by uncertainty, complexity, and increasingly nuanced relationships "strategic environments." Crucially, the strategic environment is not a place. Instead, it is defined by the U.S. Army Training and Doctrine Command as "a set of global circumstances, conditions and influences" that make up a world of complex chaotic factors in which threats to national security develop.[37] The unknown and only imaginable here becomes a site whose conditions must be opened up, captured, and harnessed into military approaches and logics. The concept of the strategic environment is not limited to the U.S. military; it has reshaped thinking in a broad array of fields, from ecosystems science to design and architecture (and, importantly, signaled the fusion of such fields to military research), and has become a generalized approach to the uncertain and unbound haze of complex circumstances that lead to future threats.[38]

According to this logic, the disastrous uncertainty in other emergent systems (such as turbulent weather, destructive wind patterns, and rising sea levels) cannot be controlled, but the rhythms of their agitations can be identified and used. As James Corner argues, the self-generating energy supplied by such events can "be guided, harnessed, siphoned and steered."[39] The trick is to find the rhythms of their dangerous pulses, to ride the waves of their risk, to locate their force fields, and to move alongside them. This perspective—one that sees emergent catastrophe as also overflowing with promise—is in perfect keeping with the project of preemptive microbial governance. The aim is not to stop the threat of emerging microbes, but merely to siphon and steer its energy so that the project of preemptive biopreparedness can meet more creative and viable ends.[40] In this context, the risk map transforms uncertainty itself into an operable problematic. Akin to the strategic environment,[41] these risk maps bring the darkly imagined future into being as a canvas for preparedness. The quality of unyielding uncertainty is rendered not as an epistemological problem, but as a challenge and opportunity for developing and sharpening technique. The limits of knowledge do not pose an insurmountable problem; instead, that very aporia becomes a generative challenge of moving, adapting, and acting in the face of open-ended anticipation.

The Depthless Depths of Risk

What is at the other end, one wonders, of those depthless patches of greens and blues that compose the risk maps? When, if ever, will the mystery termed "risk"—an enigma that the map simultaneously conceals and conjures—recede and vanish? These cartographies of threat function not to put a definitive end to the indeterminacies of risk, but to locate that space of uncertainty and hold it open in order to find its perpetually deepening forms. We see, for example, this literal stretching out of the space of the possible in the risk map's designation of threat as neither an issue of presence or absence, but as the multiplication of degree. As with other risk technologies (the financial risk score, the tracking of mobilities via border security data), animal sentinel media deploy a deluge of calculations, measurements, numbers, and other quantifying techniques to increase what Amoore describes as the "fractionation of ever-more finite categories of life—*degrees* of safe and unsafe, vulnerable and durable."[42]

Stretching risk in the index of endless potential threat redefines the relation between vision, information, and knowledge. Such recodings are significantly marked through the cultures of data. Orit Halpern contends that in a data-driven world, vision itself had been reformatted and replaced by "communication." Yet "communication" too has ceased to be understood as the conveyance of meaning. Instead, it has come to encompass the ceaseless flow of information processing.[43] This reformatting of vision also restructures the eye, which must always be available to recalibrate, to find new patterns in future information, "to anticipate and assimilate more data."[44] Accordingly, for Halpern, "credibility" too no longer posits a relation to knowledge, but is premised on having "capacity."[45] These shifts maintain the work of risk technologies (like animal sentinel media) that operate to mine data, calculate, account, process, and spread out futures in an array of open possibilities. The logic of risk technologies through which this information is generated matches the depthlessness of the object it pursues (risk), because the results it generates can never arrive at a point of completion. In contrast to the longer genealogy of visualization technologies that index in order to arrive at a point of certainty, the index of risk moves through the visual rhetoric of calculation, data, informatics, and quantification to show uncertainty. In this sense, they are technologies of uncertainty.

Do risk technologies simply not work well? Do they fail to accomplish their task? No; arriving at indeterminacy is precisely their function. They work by pointing to the emptiness of the information that is missing. They inhere in the blank spots of data, embracing unknown futures and seeking their unendingly manifold edges in an always-incomplete data flow. The point is not to arrive at certainty, but rather to reveal a multiplicity of emergent possibilities.[46] Like other risk technologies, animal sentinel media work on the premise that data's blank field points to the incipient contours of the future rather than what is already known and processed. They succeed even in—or perhaps because of—their failure to arrive at an account of something that can crystallize and become determinate, and without ever seeking to confront that aporia.[47] These technologies throw numbers and quantifications into the representational gap unreached by the usual visualizing techniques. Uncertainty is the operational space of this technology, and this uncertainty keeps animal sentinel media at work mining data. Their infinite task highlights the paradoxical nature of the closed equation in which the accounted indexical guarantee and the unaccounted risk refer back to one another. In this way, the absent or unaccountable becomes an integral part of the indexical guarantee. In stark contrast to the technologies of certainty, the index produced by animal sentinel media presents uncertainty and nonknowledge both as a site for securitization, and as the object to be apprehended through further and further calculation. It marks what is incalculable about risk while absorbing contingency itself as a basis for action and a form of knowledge.

This epistemological contortion has material consequences. The futures to which the index of risk point are always already ruined, and so they demand and remain perpetually open to anticipatory remediations through technological interventions, design, and engineering.[48] As with the technique of drawing energy (and resources) from the strategic environment, perpetual uncertainty becomes a realm of possibility in which to develop techniques and technologies shaped by the unknown, to assimilate its ambiguities, and to learn from its complexities in order to be able to move alongside ever-impending crises. These data-driven risk technologies work to continually mine the edge of the future, coaxing deepening forms of uncertainty into view and proliferating sites, objects, and networks in need of securing. Their aim is not to pinpoint, resolve, or progress from a problem, but to make the future *resilient* through the building out of global systems and infrastructure.[49] Unlike strategies of prevention, which assume that applying knowledge can avert an event, resilience presumes insufficient knowledge.[50] The goal of resilience is simply to keep a finger placed on the pulse of possible shifts in a world made up of unpredictable, autopoietic emergent crises. These crises are made operative and productive by proliferating infrastructure and technologies that refashion a world "alive with data" into a communicative orb. As Halpern and Robert Mitchell argue, resilience seeks to build a planet capable of supporting the ubiquitous informational infrastructures through which futures are now created and managed.[51] This mandate to fashion the world into a "computational planet"[52] does not stabilize risk or slow it down. It reveals an expanding realm of possible correlations between possible conduits, arrayed across a hypothetical and unspecified future time without trying to give closure to it.

In tracking the index of animal sentinel media, I have asked after the mechanisms through which these preemptive data visualization technologies bring forward their knowledge claims.

If animal sentinel media act as inscription devices that assure us of the existence of microbial risk, indeterminacy and uncertainty nonetheless return as the structuring absence of the visible indexes they produce. In converting the possible into an actionable and generative security environment, uncertainty is transformed into a changeable and workable object. The definition of knowledge mutates to enfold the nonexistent and yet-to-come. If knowing something can be defined as the condition of understanding, analyzing, and grasping it as true (and of its being true), that is not how things appear in the knowledge systems of preemption. The index of data that points to "possibility itself" now appears as a quantified quality. Knowledge no longer derives value solely from its ability to stabilize information that eradicates the unknown, but also from the guarantee of an open-ended process that promises never to end. This reconfigured form of knowledge and its elusive base belong to the early twenty-first century as an epistemological form in which absence, the imagined, and uncertainty take shape as a kind of knowledge.

Once the calculative imaginary prioritizes uncertainty as an optical trope with epistemological value, it can also be operationalized as a generative territory that must be continually plumbed. Understanding how the index of risk works can begin to bring into view the ways in which the world is reorganized under the epistemologies of preemption. This uncertain logic works to reshape the present in the image of imagination and the fictional, and renders normal the exploration of all manner of speculative technologies—from the fantastical drones that some tech experts hoped would be capable of assassinating viruses by 2025 to the scientifically reasoned manufacturing of speculative vaccines stockpiled in the event of a catastrophic flu pandemic. This logic also made it possible for scientists in 2012 to engineer into existence a highly pathogenic form of avian flu—precisely what these animal sentinel media claim to forestall and block.[53] Such moves are profoundly linked with the technological mandate that seeks to make our futures resilient by rendering the planet communicative. On both registers, the mutant epistemologies of preemptive biopreparedness call for building out the world in the form of speculative modalities. In line with the endless temporal imperative of the counterterror state, these strategies ensure that emergent microbial futures self-propagate endlessly. This chapter has examined the relationship between the optical, technical, and epistemological dimensions of speculation. I turn now, then, to examine how this technological mandate to reformat the planet as resolving technology transforms the regenerative capacities of emergent microbial risk to be transformed into a productive and endless frontier for a new and inexhaustible economy in bioinformation.

4 Fluid Economies of Biosecurity

This scene takes place in Silicon Valley. It may be midmorning in Cupertino. Figures in hoodies and sneakers stream into a building, its glass wall rising like a cresting wave. While clusters of tech workers veer toward cappuccino stations in the atrium before drifting off to their offices, a smaller team has been working overnight. In a large, open-plan room littered with coffee cups and takeout containers, many of them have been hunched over their laptops, monitoring Twitter and Google for hours. Others are scooting between data stations on careering office chairs, mashing and analyzing real-time data coming in from Chad and Cameroon. Another group has been scribbling across white boards, now layered like palimpsests, while fielding calls from Nigeria. Television screens line the room, displaying images from local stations in dozens of locations around the world. Nothing of note is happening in those remote elsewheres, but a small station in Lagos reports that several people have died from an unnamed illness in a nearby town. The team in California is working with bioinformatics engineers to deploy computer algorithms that can identify the virus and determine whether it's a threat to other countries. A data analyst processes constant and real-time flow of information on global databases, while their teammates at the central African headquarters in Yaoundé have been on the phone with local clinics for hours. Lab results are still coming in, but the virus is unidentifiable—it is not malaria, Marburg, or Ebola. The Californian team, in turn, opens up a repository data feed with its partners in Cameroon and will soon upload reams of new genetic data from specimens under examination. A room controller surveys the entire scene, while commanding five active computer screens to determine whether Nigeria needs to be put on high alert or if Dubai is now a zone of specific risk.

My evocation of this Silicon Valley control center, including the operations, countries, actors, and technologies named, is based on the scenario promoted by Nathan Wolfe, a scientist-entrepreneur praised as a pioneer in the war against microbial emergence.[1] The central command room represents his vision for a "planetary immune system":[2] a real-time global surveillance system that uses the power of computation and data analytics to halt biological threats at their point of origin. Wolfe developed an interest in emerging microbes while working as a Stanford microbiologist researching AIDS in Cameroon. There, he met communities of hunters who kill and consume local wildlife. He noted that the process of butchering requires close and sustained contact with animal bodily fluids. Wolfe became preoccupied with the bushmeat/hunter relationship as a likely site for the endless emergence of novel microbes. As a self-proclaimed "virus hunter," he set his sights on the ambitious task of discovering and cataloguing all of the viruses in the world.[3] Wolfe would attempt to do so by collecting and sequencing blood from these bushmeat hunters and their kill, and by using such populations as sentinels through which he could "eavesdrop" on "viral chatter," his own term coined to describe the constant genetic exchanges between viruses and their ecologies.[4] This approach, according to Wolfe, would enable the tracking of microbial emergence in its moment of genesis. He repurposed the AIDS laboratory in Yaoundé as a collection hub, a sequencing laboratory, and a data-sharing center, and then scaled that model up, opening field sites across Asia and Africa.[5] Lab researchers would work to collect blood samples, sequence microbes discovered there, and report and store all the resulting data.[6] This organization became the Global Viral Forecasting Initiative (GVFI, 2007–2019).[7] The information generated and shared through GVFI, Wolfe hoped, would one day serve as the backbone of a planetary immune system exactly like the one evoked earlier.

Wolfe's dream has been shared by many. From the 1990s onward, the United States redirected its Cold War fantasy of global tactical control via electronic surveillance into a program of developing high-tech total systems for the early detection of microbial emergence.[8] Such technologies are also a central part of the American dream of pandemic countermeasures. U.S. government bodies working across defense (e.g., the DOD), health and medicine (e.g., the NIH), and international development (e.g., the U.S. Agency of International Development) have supported Wolfe's GVFI research through grants and partnerships. GVFI's research grew out of the Institutes of Medicine's (IOM) call to amplify the Global Outbreak Alert and Response Network, and a series of public-private partnerships through which the U.S. government collaborated with corporations such as Google to develop global pandemic surveillance systems—most notably the International System for Total Early Disease Detection (INSTEDD), and Pandemic Preparedness for Global Health Security (PREDICT). It has also been fleshed out through transnational and international corporations, assemblies, and organizations, like private investment corporations and the World Bank. All of these institutions have supported a vision of total global pandemic surveillance, widely regarded by experts as the best and perhaps only hope the world has of getting ahead of the endless curve of emerging microbes.

It makes a certain kind of sense to examine GVFI, and the ambitions of Nathan Wolfe, from a critical biosecurity studies perspective. GVFI's mission to "stop it there before it gets here" fits comfortably within the standard biosecurity paradigms of defining and localizing risk, asserting and guarding boundaries, and immobilizing movement. Scholarship in critical biosecurity studies tends to focus on the ways in which biosecurity efforts organize the world into dichotomies (external/internal, contaminated/at risk) and the techniques of separation that work to keep bodies, spaces, processes, and objects apart.[9] In examining the politics of separation and containment, this work illuminates crucial material,

Figure 4.1. Screen capture of Global Viral Forecasting Initiative's logo. globalviral.org, 2016.

cultural, and political formations that surface through biosecurity practices. Yet to analyze Wolfe and GVFI (and, by extension, the U.S.-led war against emerging pathogens) exclusively within this framework would be to miss the fact that, in its efforts toward deterrence, GVFI courted precisely the conditions that biosecurity projects seek to avert. Rather than erecting and enforcing borders, GVFI encouraged contact and exchange across the very systems that biosecurity aims to separate (infected/uncontaminated, animal/human, wild/domestic). Instead of strategies of cessation, it mobilized and facilitated movement of resources across distances, species, states, and organizations; Wolfe's ideal Silicon Valley scenario of global surveillance trafficked in an intense range of flows and mobilities. GVFI's old website logo[10]—a graphic representation of planet earth networked together by crisscrossing lines accompanied by the words "ecology | biodiversity | public health"—is suggestive of the project's investment in situating world health in a dynamic and entangled scene (Figure 4.1).

To bring such dynamics into view, this chapter shifts away from a concept of biosecurity determined by the logics of bordering, containment, and cessation and toward one structured by the movements required of the contemporary bioeconomy. In chapters 2 and 3, I analyzed how the informational turn in the genetic sciences and Cold War imaginaries helped to reformat microbes as perpetually mutating and endlessly emerging forms of naturally occurring information. Here, I show how this view of microbial emergence as a self-rejuvenating fount of bioinformational plentitude was taken up in a U.S.-led biosecurity project that sought to fashion an illimitable economy from a world ostensibly overflowing with data. A project like Wolfe's planetary immune system requires information to appear as a powerful and flexible object, and under microbial resolution, information becomes the medium of a new global project. Information is taken to be capable of securing the world from emerging microbes, functioning as an equalizing medium in a new paradigm of international development, and enabling so-called developing nations to meet new standards of global health in ways that serve the needs of the "first world."[11] This project reshapes the world into a platform capable of opening up access to new sources of information, facilitating data harvesting and generation, and enabling the circulation of data in and through the virtual and material networks of global exchange (bodies, labs, ecologies, computers, databases, for example) such that it can become an object of economic significance. In this chapter, I use the term *fluid economies of biosecurity* to describe those processes and to name the flows called up in this paradigm of biosecurity.

By directing attention to biosecurity flows, I trace how the U.S. war against microbes pursues two simultaneous and interlocked goals: first, to secure the world from emerging pathogens via a global health project; and second, to enable an economically uneven world to "find its own level"—that is, to magically level itself off—as a natural consequence of the fluid economies of biosecurity. In this sense, the project of microbial resolution operates as a dual-purpose endeavor: it serves a set of overarching, opposing priorities while massaging out their contradictions to make them one. This articulation of global health yokes together the goals of capitalism and humanitarianism, militarization and global health, the economic and the biological/biospheric in ways that make their disparate threads and aims coterminous. To capture these contradictions, I deploy the term *resolution* in this chapter in multiple senses: as the solving of conflict, as the solution to a problem, as the expression of will of a governing body or assemblies, as the process of making commensurate, and as a series of discordant notes brought into harmony. I hold these meanings in mind as I examine the processes and promises at work in the fluid economies of biosecurity.

In many ways, this chapter is a story about how and why information itself became such a central object in the U.S.-led war against emerging microbes. This matter cannot be understood without first taking into account the relationship between the contemporary bioeconomy and the U.S. economic crisis of the 1970s. In the 1970s, it became clear to many observers that the material resources upon which the U.S. postwar economy relied were quickly being exhausted, thus threatening to impose limits on the U.S. economic future. I situate the rise of the innovation economy in this period, tracing out how economists and politicians worked to bypass finite material resources by designing an economic model built to capture the ceaseless property of change itself. Based on my previous discussion of microbes as autopoietic systems, I show why information, and bioinformation above all, came to be the core object through which this economy would operate. If the innovation economy was designed to enclose the property of endless transformation itself, information was the object through which that endless potential could be accessed. This was an integral part of the neoliberal restructuring of nature as value. In their transmutation into information, autopoietic natural systems like microbes—and the boundless immaterial resources and futures they purportedly represent—became subject to the fluid economies of biosecurity.

In the name of biosecurity, a data-driven global pandemic surveillance system must unite a world of different actors with divergent interests under a common purpose and mission: states, NGOs, individuals such as Wolfe's bushmeat hunters, his DOD and NIH backers, his Cameroon-to-Palo Alto collaborators, corporations, academies. I analyze the implications and ramifications of this effort, demonstrating how the United States deployed what I term the *language of harmonization* to promote the idea that nations should overcome various differences to meet the common threat of emerging microbes. The language of harmonization was not a platitudinous gesture toward global peace, inciting the world to "join hands in perfect harmony," nor did it operate under that pretense. Instead, it emerged as a regulatory discourse of commensuration. I find this discourse laid out across a constellation of grey literatures from American government boardrooms: in global resolutions, policy documents, reports, conventions, mission statements, and treaties. The language of harmony acknowledges that nations possess unequal economic, scientific, and technological capacities, and seeks to even out that mottled global landscape. Only by leveling the operational field of development in science and technology could visions like Wolfe's planetary immune system be made possible. I analyze how this discourse of harmonization operated in the global regulatory environment around the handling of bioinformation. Often ignored, this monotone discourse gets relegated to the boring, fluorescent-tube-lit realm of bureaucracy. I read this body of grey literature as itself a *resolving medium* through which the world is rewritten as a regulatory environment that gives information legibility, accords it a spectrum of values, and maintains it in an unyielding field of economic relevance. If one aim of this book is to track lines of vision and mediation at the point where emergence discourse presses them to their limit, here vision and mediation are reconstituted as bioinformation from within the bureaucratic grey space of global regulations.

In calling upon all nations to come into the fold of new standards of global health, the language of harmonization works to produce, fashion, and regulate the world for the circulation of bioinformation. I examine how these transformations in turn render the world in a universal format—a vast protocol space or operating system—capable of handling the logistical orchestration of flows of goods as bioinformation around the globe. This reformatting refreshes the innovation economy, while making bioinformation the key object for health security and for U.S. economic, imperial, and military imperatives. Reshaping the transnatural biosphere into a vast medium of information transmission is part of a geospatial strategy of territorial expansion that rearticulates and expands the reach of the United States, and other Western powers, with their militarized logics in a system that constantly renovates itself. While critical scholarship in biosecurity provides invaluable ways to understand health and militarization, a focus on the security techniques of cessation and containment has led scholars to read the war against microbial emergence as productive of its paradoxical opposite: a state of insecurity.[12] However, when we read the war against emerging microbes as a project that traffics in a range of flows, this production of insecurity no longer appears bracketed off as an unforeseen and unintended consequence, an exceptional accident within a larger system that seeks to increase security. The production of insecurity is not a paradox, but remains wholly consistent with a broader and more complex logic at the heart of the project to govern and control microbial emergence.

From Resource Exhaustion to the Endless Frontier of Innovation

In the decades following World War II, the United States was clearly the main driver of the international economy. The emergent Bretton Woods system rewrote world relations according to global finance and creditor/debtor structures to craft new revenue streams. The Marshall Plan, in turn, spread American influence throughout western Europe while securing those countries within systems of Cold War economic expansion and aid. Yet between the late 1960s and the mid-1970s, the United States had to navigate a period of dramatic economic decline. This downturn was rooted, at least in part, in the environmental crisis. Signs of planetary stress were everywhere, and the rise of environmentalist movements brought issues of pollution to public attention throughout the 1960s. The 1962 publication of Rachel Carson's *Silent Spring,* among many other markers, directly connected the use of agrochemicals (specifically DDT) to metabolic shifts in the earth systems that ultimately led to species extinction. Climate science provided new ways of measuring and quantifying rising CO_2 levels in the atmosphere.[13] Amid these apocalyptic prospects, the survival of the earth—and of all life on it—came into question for the first time. Resource scarcity became a lived reality with the "shortages" and energy crisis of the oil embargos of 1973 and 1979, a concurrent rise in the cost of manufacturing, and a worldwide food crisis most visible in the famines that plagued several Asian and African countries. Together, these crises indicated that humans were already living beyond the earth's geochemical affordances, and U.S. economists were awakening to the fact that Fordist production was quickly outpacing the earth's.[14] An uneasy belief took hold in many quarters that the American economy would soon run out of the raw resources on which it relied.

Many believed the depletion of these material resources to be irreversible. When the Club of Rome published their influential 1972 report, *The Limits to Growth,* there was nothing particularly new about its projections of biospheric collapse.[15] Nevertheless, that document mobilized the concept of exponential systems to explain why this destruction could be slowed, but not ultimately halted. The exponential growth of human populations, it argued, would lead to a parallel increase in demand for raw resources such as minerals, oil, land, and food. These factors would place pressure on manufacturing industries to intensify production, especially in those sectors largely responsible for biospheric catastrophe through chemical pollution, rising CO_2 levels, and deforestation. Deploying a systems approach, the report illustrated and underlined how this system would produce "negative feedback loops," ensuring a precipitous descent into crisis and catastrophe.[16] With the earth in a dire state of disequilibria, the wide horizon of economic opportunity once open to the United States was quickly contracting;[17] American economists viewed the planet's material exhaustion not only as a crisis for the future of life on earth, but also as an irreversible and catastrophic limit to economic growth.[18]

To respond to these impending limits to economic growth, theorists of the New Right in the 1970s argued that the United States would need to navigate away from the world of heavy manufacturing and to divest itself of a world of finite material resources. These theorists put forward a new economic model built on resilient, abundant, and inexhaustible resources. As neoliberal political philosophies rapidly gained ground with the New Right from the late 1970s onward, Thatcherism in Britain and Reaganism in the United States initiated a radical reshaping of the global economy. Unmooring economic production from Keynesian paradigms of regulation and welfare, these processes instigated specific measures to unleash the forces of the free market in pursuit of exponential and unimpeded growth.[19] The protections and obligations that defined the welfare state in the Keynesian obligation model—and that functioned, among other things, to keep the economic and the biospheric separate and distinct—would be erased and replaced. A crucial dimension of this neoliberal reorganization of life, as Melinda Cooper argues, "lay in its intent to blur the boundaries between the spheres of production and reproduction, labor and life, the market and living tissues."[20] Neoliberal economists would see in the new worlds formed at the points of their conflation the promise of a new kind of resource that was highly renewable and, for this very reason, capable of unbinding production from the limits imposed by material realities.[21] Further, by radically restructuring the economy in this way, the United States could assert its independence from the material and political circumstances of foreign countries and stabilize its leadership over the international order.

The new economy would be rebuilt between the late 1970s and 1990s around the search for an illimitable resource. While the American economic system was in decline, competitors such as West Germany and Japan were experiencing favorable growth. From the 1976 National Science and Technology Policy, Organization and Priorities Act onward, it was argued that the economic ascendancy of these other countries was taking business and profits away from American industry. The root of the problem, that bill contended, was an "innovation crisis" across the nation.[22] Concerns over the status of American innovation persisted in government discussions throughout the latter half of the 1970s in a series of surveys and commissions dedicated to the topic. The U.S. innovation economy was largely formed in the wake of these debates. Across such studies, discussions, roundtables, and meetings, the so-called crisis in innovation was repeatedly attributed to a system that kept government-funded science trapped in the public domain. Many asserted that regulation had the "effect of retarding technological innovation or opportunities for its utilization."[23] In order to support American innovation, these reports asserted, the government must create a hospitable environment in which innovators could thrive. Amplifying the already extensive reach and resonance of the military-industrial complex, those discussions recommended denser alliances between science, technology, research and development, and industry. Building on the technological origin story that views scientific, technological, and economic expansion as inherent American rights and properties, this discourse framed such elements as the very heart of the nation's being and purpose. Through innovation, the world itself could be written over as a "second creation" in harmony with the first: "God's creation" and "manifest destiny" are virtually one and the same in this ideological discourse.[24] Nothing less than the fate of the United States—both its global dominance and its ability to grasp its own history and future global mission—was at stake in the resolution of the innovation problem.

To restore the "pioneering" characteristic that made the United States the most highly advanced industrial nation in the world, it would need to operationalize the frontiers of constant change. This should be done, the New Right argued, by commercializing publicly funded research, largely carried out in university laboratories, and to incentivize this process through a radical extension of patent coverage.[25] That development culminated in the 1980 Bayh-Dole Act, which yoked together the realms of private industry and public research and built up the innovation economy at their interface.[26] It did so by making it possible, indeed almost mandatory, for publicly funded research to be channeled into the corporate realm by granting incentives for researchers to enclose their discoveries within the protective armor of patents. I return to Bayh-Dole and the topic of patent extensions later; here, I want to note that the main objective of the act was not to capitalize on any inventions, but to draw on the promise that corporate-sponsored research would lead to the development of new products in the future. The Bayh-Dole Act, in this way, operated according to a speculative logic: its goal was to keep product pipelines teeming with promising research and to enclose that promise within systems of property, even in the absence of any apparent utility, identifiable outcome, or defined end-product of a given work. As a result, one critical effect of the act was to redefine "innovation" not as "invention," but instead as a process that worked on the perpetually incipient edge of transformation.[27] In relocating economic production within the intangible process of change itself, this new economy would not be restricted by the finitude, slowness, and difficulty of material resources.

But what was the object of this new economy? Whereas material resources—steel, minerals, oil—surfaced as key resources of the era of Fordist manufacture, "information" formed the basis of the emerging system. The innovation economy was born from a period in which rapid developments in information-based technologies—the invention of the first commercial barcode, the first optical fibers, the first cellular phone, the first home computer—transformed the ways in which information could be generated, processed, stored, and transmitted. It was fueled, as well, by the rise of the information discourse of the 1960s and 1970s, which drew from the digital utopianism of California counterculture by importing the value of "one world/one mind." Centering on the cybernetic mind as the site of social change, maverick counterculturalists (who later formed Silicon Valley) believed that, behind the physical world, hidden streams of information emanated from everything and everywhere, looping in circuits and constantly generating new patterns.[28] From the 1960s onwards, all these factors contributed to fashioning a potent informational imaginary that would have profound consequences. In this imaginary, information was regarded in deeply optimistic terms, because the future could and demanded to be reshaped through the multiple powers and textures of data itself.

Bioinformation, or a New Medium for a New Economy

While the innovation economy dealt in all manner of informational products (CDs, telecommunications, software), bioinformation formed the basis on which it was built.[29] In chapter 2, I analyzed how the rise of theories of nature as essentially autopoietic systems complimented theories of the inherently informational and computational logic of life. Such theories postulated that when a biological entity such as a virus meets a threat to its existence, its systems self-organize to find, within their own limits, a catalyst for self-regeneration into new and ever-complicating forms. Life systems are consequently viewed as perpetually disruptive systems that innovate from a point of crisis. In keeping with the neoliberal rhetoric of calamity-born capital, from the 1970s through to the 1990s, theories of complex biological growth were installed within those emerging narratives of economic development.[30] The understanding of life as crisis-born systems of constant, self-organized transformation made for a rich foundation on which the United States could construct new strategies of economic production and resilience. Meanwhile, the reformatting of genes as data enabled the exploitation of biospheric processes as informational, and thus subject to economic processes. In this way, the innovation economy was built on the hope of making use of nature's own process of endless mutability.

In contrast to Marx's concept of production as a process through which nature is transformed by human labor, the innovation economy therefore apprehended life systems themselves as a form of natural labor, expanding the very terms that defined the economic process.[31] No longer did the productivity of human labor create value from raw resources; the innovation economy fixed on the capacity of life to make itself perpetually anew,[32] and sought to access its essence in seemingly immaterial and inexhaustible bioinformation. By isolating the data of bioinformation, the workings of nature could be understood and seized.[33] Collecting bioinformation became synonymous with owning production in concentrate. This distillation of an object to its essence, in Bruno Latour's terms, was a way of "having as much and as little of something as possible." It offered a means of capturing inconvenient matter in more transmissible forms.[34] Such tension "between [the] presence [of the original/whole specimen] and [its] absence," Latour argues, is what we call information.[35] By isolating information, we could have "the form of something without the thing itself."[36] Rather than collecting material specimens per se—the plants, rocks, insects, or tissues of past colonial expeditions—the goal of the innovation economy was to collect and sequence their bioinformation. Once biological matter was flattened and expanded as information, it could be shared, recombined, modified, isolated, acted upon, and replicated infinitely to add and increase value.[37] For example, one might obtain a soil sample, isolate a microbe within it, sequence its genes, identify its genetic code, and store this information on a database for some undetermined future use. As Bronwyn Parry states, this process conceptually disaggregated information from the material from which it was derived, transforming "genetic matter" into a "genetic resource" to be monetized.[38] In collecting bioinformation, then, the undiluted forces of production itself could be extracted and collected as a raw resource without the constraint and threat of material limits.

And while nature-as-bioinformation existed across the globe, advocates of the innovation economy believed its richest sources were to be found in developing countries. In this view, the nature of wealthy, developed nations was less bountiful, long since uncovered, and already exhausted. In contrast, undeveloped countries were rich in untouched nature that teemed with diverse genetic material waiting to be discovered. Heavily inflected with the expansionist fantasies of the Cold War, the space race, the American West, and other quests for new worlds, the search for genetic resources approached these territories as terra incognita, as endless frontiers of bioinformation awaiting the interventions of Western science and technology.[39] By this logic, leaving nature locked away in third-world countries amounted to wasting future productivity. How to unlock the forces of undiluted innovation and boundless production from the natural resources of the third world? The history of resource reallocation from the Global South to the North is an old saga of geoeconomic power via biocolonialism.[40] Indeed, those networks of extraction came under intense scrutiny in the 1980s when activists condemned them as acts of biopiracy, direct extensions of colonial science through which the West continued to profit from the non-West. This threw the bioprospecting of U.S. scientists and industries into public view as a highly contentious activity. Still, U.S. scientists and industry leaders were convinced that the best nature was in so-called gene rich/cash poor countries, countries that had quickly amassed insurmountable debts to the World Bank and the IMF after the Second World War. In order to meet the economic demands put upon them through such institutions, governments in low- and middle-income countries were more inclined to exploit and sell their resources through practices such as the decimation of forests for lumber.[41]

The geopolitical and economic imperatives of wealthy Western states—to render the colonial dynamics of resource extraction clean and legitimate—were therefore baked into the production of biodiversity. Beginning in the 1980s, the United States and U.S.-led institutions (pharmaceutical companies, as well as health and science bodies) began building biodiversity prospecting agreements to redistribute value between the North and South.[42] Meanwhile, the 1987 Brundtland Commission report, *Our Common Future,* asserted that postwar development models were failing: not only did such programs prove ineffective in ameliorating the economic situation of third-world countries; they deepened inequity while destroying the planet. In line with the neoliberal beliefs of Norwegian prime minister Gro Harlem Brundtland (who would become the WHO's director in 1998), the commissioner of the report sought to redress this imbalance by advocating for "sustainable development": the strategic stewardship of nature in pursuit of economic development via economic globalization and neoliberalization.[43]

By the early 1990s, biodiversity was a centerpiece of global and economic discourse.[44] The United States assumed a leading role in shaping the global discourse on biodiversity and conservation (crucially, they did not ultimately sign onto it—a point I discuss below). The meeting of the United Nations to form and ratify the 1992 Convention on Biological Diversity (CBD) further legitimized the process of using earth's nature as a new resource frontier and bedrock of a new developmental paradigm by formalizing such systems of geoeconomic power into international legislation.[45] These constructions of biodiversity were important in helping to stabilize the dynamic and mutant forces of productivity found in natural systems into the flexible form of information capable of laundering the material conditions of its own extraction. In this way, biodiversity serves the economic and ideological needs of Western nations in the bioeconomy. In order to truly do so, however, the concept of biodiversity must operate on a series of productive inconsistencies, and it is to these that I now turn.

The Dual Natures of Biodiversity

Although biodiversity often appears in the cultural imaginary as part of a positive global effort to protect planetary well-being, that discourse must be understood in all its ambiguous complexity. The project of conserving biodiversity is simultaneously a call to ecological rebalancing, a new model of economic redistribution, and a mechanism to launder the process of resource acquisition.[46] Not only is it a strategy to enable "technology-rich/resource poor" nations to obtain material from "resource rich/technology poor" states; "conservation" is also "future building," a project to grasp and apprehend biological resources across space, and—more crucially—to secure worldwide resources through time. In addition to protecting the riches of nature found in "resource-rich/technology poor" countries, biodiversity conservation frames genetic diversity as a realm of perpetually undiscovered unknowns in nature. As Cori Hayden remarks, in biodiversity, it's less important to know what we don't yet have as it is to imagine what we do not yet know we have.[47] In bioeconomic terms, this unknowability is crucial to making a boundless world of perpetually undefined value. The authors of the 1992 *Global Biodiversity Strategy,* to cite one example, frame nature's undiscovered diversity as a boundless frontier of innovation, and thus as the centerpiece of a new and resilient economy:

> The unknown potential of genes, species and ecosystems represents a never-ending biological frontier of inestimable but certainly high value. Genetic diversity will enable breeders to tailor crops to new climatic conditions. Earth's biota, a biochemical laboratory unmatched for its size and innovation, hold the still secret cures for emerging diseases. A diverse array of genes, species and ecosystems is a resource that can be tapped as humanity's needs and demands change.[48]

Rather than diverse nature per se, here the "unknown potential of genes, species, and ecosystems" becomes a profoundly flexible and vast resource. In this vein, biodiversity discourse frames nature as an endless storehouse of unknown and untapped bioinformation, a vast reservoir of inestimable potential value that as Hayden argues, "can only be imagined,"[49] and from which future biotechnologies promise to arise.[50] In this way, biodiversity discourse transmutes life systems into informational resources. Biodiversity conservation thus grants specific actors access to earth's riches in ways that enable them to index and maintain hold of the always secret frontiers of inestimable future value. Failure to protect biodiversity would be experienced in the economy as informational loss and degradation, the vanishing of untold potential future value.[51]

Crucially, microbes have long been precluded from such ownership agreements since research on pathogens generally proceeds from a scientific ethos of open sharing of virus and data; it is generally agreed that the unfettered dissemination of this material is in the interest of the global public.[52] While the provisions of the CBD maintained this exclusion of microbes, it nevertheless performed powerful cultural work in its reframing of life systems *on the whole* as informational *in essence*. Therefore, they also became open to economic processes in the information economy. Microbes normally go through a process of acquisition, sequencing, and research and development; assertions of ownership can only be made from the point at which they are materially transformed into genetic sequences and onwards.[53] The CBD complicates this: the unownable status of microbes was not changed by this agreement, yet its framing of nature-as-information opened up possibilities for new ways to claim ownership of microbes. Indeed, in 2006, then Indonesian health minister Siti Fadilah Supari invoked the CBD to assert Indonesia's "viral sovereignty"—the claim that viruses originating from within a nation are part of their national patrimony—over avian flu virus samples that it had been sharing with developed nations. Supari did this in protest against what Indonesia (and many other nonhegemonic countries) regarded as a neocolonial dynamic of the global virus sharing economy: the most economically powerful nations and corporations would use viruses shared by disadvantaged nations to make profitable biotechnologies (such as vaccines and antivirals) from them, while sharing nations would be those least likely to benefit from any of these in the event of an actual pandemic.[54] As Natalie Porter and Amy Hinterberger argue, ownership claims like Indonesia's seek to restructure value in the global microbe sharing economy by "tethering" microbes to specific times and places.[55] While such efforts led the WHO to devise international agreements (notably, the 2010 Pandemic Influenza Preparedness Framework)[56] to address these inequities, the CBD's framing of nature-as-information still had powerful effects on how microbes could be treated. Once understood as inherently informational, microbes became speculative and promissory objects, making unclear the point at which their value begins to accrue. The abstraction of microbes into information also permits their complex journeys over global databases that obscure the lines that connect microbes to value claims based in time and territory. Even amid efforts to make viral sharing more equitable, then, the valuation of nature as/for bioinformation became a mechanism through which value claims on viruses could be inserted at multiplying points within the evolution of a given pathogen, even in some cases rendering its material status so ambiguous that some actors came to understand value as inhering at the moment of its instantiation in an outbreak event.[57] And so while the CBD did not change the ownership status of *microbes as such,* its construction of nature-as-bioinformation opens up and activates numerous possibilities and processes by which value may be drawn from pathogens, meanwhile rendering unclear and expanding the points at which claims to patentability could be inserted. Despite precluding pathogens, the CBD nevertheless retained its monetizing and laundering effects in that realm.

As Hayden notes, biodiversity, across its various inflections, links the mandate to protect the earth's natural resources with the economic imperative to guard not merely against a loss of profit, but against all limits to future economic growth. Biodiversity discourse, then, represents a qualitatively new relationship with nature, framing the plentitude of earth as an "ecological workhorse" and a "storehouse of valuable genetic information," as well as a resource to be stewarded "explicitly for economic" reasons.[58] Because nature labors and produces, the proponents of biodiversity conservation argue, the earth's biota must be conserved in order to bring about what both economists and conservationists term "natural capital."[59] Such language designates nature as a "stock of renewable and non-renewable resources" constituting a "trust fund" for all people. As "human beneficiaries" "live off the interest the fund provides," nature must be protected from a depletion that would diminish the "returns drawn from its dividends," or nature's "ability to continue to provide benefits over time."[60] "Natural capital"—the term is striking. We will later see the use of other such "corporate oxymorons," rhetorical sleights of hand deployed by corporations to limit critiques of their harmful impacts and define the terms of whatever scrutiny they do allow.[61] Clearly, this view of nature was crucial for those new economists who sought pathways out of an economic crisis largely rooted in resource depletion. Through biodiversity discourse, the once very different realms of economic development and ecological concern become mapped onto one another such that they are made "logical[ly] consistent." Economic, environmental, and social policies should no longer be understood as competing against one another, but should be reconciled to "yield the full benefits and logical consistency of natural capitalism."[62] Nature, once produced as a valuable source of self-regenerating raw material upon which the new innovation system could be built, would need to be protected to bring about a new and boundless economy. This dual-purpose endeavor gave the biodiversity concept an oddly inclusive and expansive hybrid—what the authors of *Natural Capitalism* might have termed "green both ways."[63]

My point is not that nature has been commodified in this process. Nor am I seeking to direct attention to any hypocrisy or duplicity in the project of biodiversity conservation. In fact, biodiversity discourse is quite clear and open about the dual nature of its effort to pursue biodiversity conservation as both a normative moral goal and a strategy for economic growth. Instead, I am examining how the unknown futures of life systems themselves are reformatted as informational resources of unlimited value when the economic is conjoined to life, making these two once divergent realms "logical[ly] consistent." Just as corporate oxymorons work to conceal harm countenanced by corporations through a reversal of figure and ground[64] (for example, "safe cigarettes," "sustainable mining"), generating "natural capital" depends on the production of "cheap nature"—the devaluation of nature as the very foundation and precondition of its value.[65] While the dual natures of the biodiversity concept rely on resolving such contradictions, that resolution is not seamless. Instead, it produces key structural and conceptual tensions. I will examine how these tensions become lodged and articulated within the workings of a global health project in the following chapter. For now, I consider the ways in which a parallel kind of oxymoronic drive is at work in U.S. efforts to rewrite the landscape of world relations in preparation for a global health security project. The project of microbial resolution requires new forms of international collaboration, but in order to fashion these, it must first address its own related series of contradictions and incommensurabilities.

The Language of Harmonization: Raising Capacity for the Information Economy

In chapter 2, I argued that the emerging microbes worldview frames the planet as a transnatural biosphere, a vast space inextricably networked and traversed by microbial tides. From the molecular to the global, across natural systems and manufactured processes, this standpoint views virtually everything in the world as united by its mutual susceptibility to microbial risk. We might call this *microbial kinship,* a sense that all things are made up of, and therefore bound together by, common biological and social substrates. Indeed, this ecological notion of kinship was accompanied by the appearance of a specific rhetoric in U.S. government literature from the 1990s onward, a discourse that circled around terms such as "common problems/common needs," and "shared risk/mutual responsibility."[66] These taglines are representative of what is now a key tenet of global health: that a world permeated with emerging microbes is a space united by microbial risk, and consequently requires a harmonized approach to the problem of emergence. Mobilizing this harmonizing discourse in global health through its influence at the WHO and International Health Regulations, U.S. experts impressed upon the global community the important role that each of its members (nations, individual actors, institutions, governments) played in the battle against emerging microbes. In advancing a plan for global health built on the universalizing fantasy of mutual stakes and the presumption of shared interests, the United States posited in the 1990s that all nations have responsibilities within a global pandemic forecasting project as members of a world community.[67] This ideological rhetoric of commonality, mutuality, sharing, and oneness invokes a vision of a world joined together in relatively equal and cooperative terms through a shared mission to combat microbial threats. Deploying the unifying terms of "common problems/common needs" along these lines, the IOM had advanced its vision for a plan to "raise the world's capacity" for global health.[68]

What did it mean to raise "capacity"? The term had very different implications for wealthy nations versus the rest of the world. From the American perspective, it had been understood as a future-building project that was both technocratic and universalizing. As a technocratic project, building capacity required the union of a broad range of actors. It was a mode of intervention that sought to proliferate technological, informational, and economic production in the future in order to render it responsive to any potential crises. These features were also present in Wolfe's vision for GVFI's planetary immune system and its precursors. The IOM described the factors involved in their version of raising capacity, which include enhancing bioinformational exchange to facilitate a range of anticipatory technologies (such as vaccine production and stockpiling, or the computational modeling of viruses), expanding remote-sensing technologies, sharpening data tracking, strengthening and networking of all manner of databases throughout the world, and advancing diagnostic tools.[69] Such measures would not halt or prevent viral catastrophes in the present, nor would they address the conditions that might lead to microbial emergence. Intersecting with the temporal orientation of biopreparedness discourse, these projects and technologies were thought to equip the United States so that it could, in theory, build a resilient future.[70] The way to address and implement this plan was to strengthen existing global partnerships and to forge and court new ones in an array of areas including pharmaceuticals, NGOs such as UNICEF or the Food and Agriculture Organization of the UN, universities, and other nation-states.[71] Such an approach would ensure that American interests in the broadest sense—corporate, political, military, scientific, and industrial—would be protected.

At its inception in the 1990s, raising capacity was also a universalizing project, and the United States envisioned itself as the lead. Working through its influence with global institutions like the World Bank and the WHO, the United States sought to craft new measures and standards, and to internationalize these in keeping with *America's Vital Interest in Global Health*.[72] One key goal of this effort was to "establish and maintain core capacities for surveillance and response" and monitor compliance by the UN's 196 member states with the new requirements to which they were legally bound through international regulations.[73] But from a Western metric, developing nations were technologically and scientifically less advanced, and so could not take up their role within this new "participatory" framework of global health. The United States therefore called on richer nations to assist in "raising capacity" so that underresourced nations could enter the universalizing fold of common measures and global standards. In order to make the fullest use of these informational resources, experts and consultants agreed that the United States should lead the global surveillance network, given its natural strengths and its industrial expertise in information and communication technologies. Noting that the global health effort faces the "special challenge" of "how to help developing countries to advance their capacities in the fields of information and communications," the authors of *America's Vital Interest* explicitly position the United States at the helm of this effort: "The United States, particularly the corporate sector, has much to offer in this enterprise." Yet in order to foster the involvement of the corporate sector in this new information-based economy, they argued that "the US government, along with its counterparts throughout the world, must ensure that the regulatory, legislative and market conditions necessary to attract private investment in telecommunications, information and technology and information services are in place."[74]

For developing nations, then, "raising capacity" entailed something quite different from the obligations of technologically advanced and wealthy countries. The government literature characterized less developed nations as not yet able to assume their role in the global pandemic surveillance network. From the perspective of U.S. politicians and experts, those countries were not only "breeding grounds" for microbial emergence; they were also lacking in scientific and technological education and training. This incapacity meant that new viruses could move undetected into the rest of the world. To "help [low- and middle-income] countries help themselves"[75] so that they might assist in the greater cause against microbial emergence, *America's Vital Interest* recommends enlisting and training people and institutions in developing countries as "participatory" collaborators, imposing measures in line with newly devised global health standards, and reopening Cold War field stations as epidemiological listening posts.[76] To that end, Lawrence Gostin, international health law specialist, recommends that international bodies like the International Health Regulations (IHR) should set "surveillance standards and monitor compliance; and facilitate small-world networks where diverse groups of scientists, health professionals, membership associations and non-governmental organizations monitor health threats" in order to serve global and American security needs.[77] The strategy of raising "capacity" here ensured that "small-world networks could transform into rich sources of information through advanced communications"[78] and would become integrated within the global information economy. If raising capacity was a future-building project for developed countries, developing countries provided the base materials.

The language of harmonization sought to loop all nations together through systems of common priorities and measures. Its call to raise capacity articulates a program in which states must address themselves to the needs of the new bioinformational economy. This project pivots on the production of information as an ideal object of immaterial and constant flow. As Deborah Cowen shows, in the capitalist schema, movement not only enables the process of accumulation by moving commodities around; circulation in itself is productive as a mechanism of value creation.[79] In the same vein, David Harvey remarks that "capitalism is value in motion."[80] Circulation spawns commodities. Bioinformation too must circulate if it is to generate the value it promises for the new economy. Yet for information to glide through space-time, the world itself must first be rendered into a universal format to enable such circulations. A global pandemic forecasting system like the one envisioned by Nathan Wolfe requires an even world: one under a unified purpose and vision, and transformed into a homogenous and universal platform capable of managing the flow of bioinformation. The question faced by those wishing to mount a global pandemic forecasting system was how to smooth the conduits of an uneven world so utterly divided by immense socioeconomic differences, and therefore also with vast gaps in technological and scientific capacity.

The very idea of making the world more even, and of "lifting up" "poor" nations, long pre-dates the rise of the bioinformational economy. In this new economy of limitless growth, information circulation replaced earlier models of postwar international development. The language of harmonization implied that seamless informational circulation alone could achieve developmental goals where prior strategies failed or fell short. The project of raising capacity sought to connect disparate states in very different economic registers by linking them within a broader information ecology. In this way, it was profoundly shaped by longstanding fantasies of virtualization in which every thing, process, and place is flattened into uniform bits of information that can move unfettered across all realms. This fantasy insists that information bears no significant relation to the conditions of its material manifestation. According to this belief, if one could apprehend the world as information alone, it would be possible to access, understand, and govern that realm in ways that transcend the difficulties of terrestrial physicality, politics, and conflict.[81] Information is figured as an inherently flexible object capable of exploiting production without the limits of material resources, thereby neatly circumventing any critiques of colonial resource appropriation. And information appears as a neutral object capable of addressing, at once, humanitarian and capitalist goals. The project to raise the world's capacity was built on the dream that information, if properly managed, would somehow bridge the yawning economic gap between the first and third worlds; by connecting the "promise of access" to a "political economy of hope," developing nations could improve their economic prospects by becoming directly connected to a global informational economy.[82] Through information, and through addressing the needs of the bioinformational economy, this project promised to bring an uneven world into commensuration with new global (read normative) technical, moral, and scientific standards. In this vision, developing nations would be connected to, and thus able to partake in, the fluid informational economies of biosecurity. For those invested in the bioeconomy, information as a medium replaces and carries out the work of development, an inherently beneficent resource whose flow alone could magically ameliorate an economically inequitable world by helping it find its own level.

Fixing Flow: TRIPS and the Control of Bioinformation

Yet for information to function as the great leveler, it must be productively and specifically channeled. Once nature was understood along a common measure as valuable bioinformation, advocates of biodiversity discourse determined that such data must be routed through the proper channels to avoid a kind of informational "rotting on the vine." Yet this issue became particularly fraught in very specific ways, through debates over intellectual property, from the late 1970s onward.[83] During the 1950s intellectual property was not yet an important topic for U.S. pharmaceutical corporations. At that time, developing countries had not industrialized to a degree that would enable them to produce biotechnologies. As nations such as India and Brazil became increasingly industrialized throughout the 1970s, however, it suddenly became possible for them to produce crucial medical goods. These countries began manufacturing the same drugs as the United States, but their products could be sold cheaply to other developing countries where the United States had once held the major market share. The accompanying loss of corporate income and profits derived, in the view of analysts and legislators, from the fact that "innovation" had not yet been sufficiently enclosed by patent laws and protections. It was hence unshielded from free use by others.[84] It became imperative, then, to direct bioinformational flow through international channels in ways that would prove to be most economically beneficial for the United States.

As I have shown, the discourse of biodiversity valued nature for its productive capacities and sought to harness its "natural labor" in bioinformation.[85] This discourse was enacted in the 1992 UN Convention on Biological Diversity (CBD), but from the perspective of the United States, the CBD lacked a bold monetizing structure. Indeed, although the United States was heavily involved in the negotiations around the CBD, it was ultimately the only major nation that did not sign on to it. Although there were many complex reasons the United States did not ratify the agreement—for instance, the Pharmaceutical Research and Manufacturers of America loudly decried most aspects of the benefit-sharing agreements—chief among these was the fact that the CBD failed to offer provisions for the control over long-term ownership of bioinformation.[86]

Lack of intellectual property protection deterred corporations from investing in the new information-based economy because there was no incentive to address market conditions of the needs of developing countries (the most populous future consumers of American biomedical commodities) that were most pressing—or so the argument went. What was required was a global regulatory environment through which interested parties could make property claims on bioinformation as a means of monetization. This process would enable the enclosure and privatization of knowledge in such a way that it would no longer be treated as a common good but instead as an ownable property. Beginning in 1986, Pfizer and IBM organized a small group of exclusively North American CEOs in the software, entertainment, and biotech industries who joined together to form a powerful private lobbyist group, the Intellectual Property Committee (IPC).[87] The IPC launched an aggressive campaign through which it conveyed to the American press, government, and public that the economic crisis in the country was the result of lost income from "unprotected" innovations. The IPC argued that, in the absence of global intellectual property laws, the United States had essentially given away knowledge and its concomitant profits to the world; put bluntly, developing countries were stealing American capital.[88] As Cooper observes, this was a preposterous claim, because something that is not owned cannot be stolen.[89] Nonetheless, the IPC worked to gain support from other major developed countries, finding allies in Western European countries, Canada, and Japan. It ultimately worked through the General Agreement on Trades and Tariffs, which governed international trade and tariffs in the postwar economy, to forge what is perhaps the most significant piece of global legislation of the twentieth century: the 1994 Trade-Related Aspect of Intellectual Property Agreement (GATT/TRIPS, or TRIPS).

The TRIPS agreement aimed to isolate and control two of the most promising knowledge and information-based industries of the twenty-first century: digital and biotechnologies.[90] TRIPS extended intellectual property rights (IPR) over a broad range of information and knowledge-based objects and processes such as trademarks, copyright, geographic indications for foodstuffs, software codes, and certain microbiological processes.[91] Under TRIPS, it was illegal for industries in developing countries to produce crucial biotechnologies. In this way, the agreement effectively maintained the division of global bioinformational ecologies by making it increasingly difficult for developing countries to advance their own information economies. Such nations, then, would continue to serve primarily as "small-world networks," crucial informational indexes of microbial risk. Biodiversity discourse operated to help these "technology-rich/resource-poor countries" to extract biological samples (most importantly in their essential form as bioinformation) from "resource-rich/technology-poor" countries, while framing that activity as part of a common investment in biological preservation. The TRIPS agreement formed what the United States viewed as the missing second part to this process: it established an international trade system rooted in intellectual property rights that enabled extended patent claims over these biological resources. These intellectual property patents are now so heavily guarded by the United States because patent recognition in developing countries is the primary means through which U.S. corporations counter the threat of "innovation extinction."[92]

Notably, then, the project to "raise the world's capacity" for global health emerged precisely when developing countries were expanding their scientific and technological capability, making the possibility of health for all at least a potential reality. These so-called late bloomers were "slow" to industrialize because they held oppressive debts incurred under the structural adjustment programs of the Bretton Woods system.[93] Only once these countries gained economic mobility and capacity were they undermined. For example, TRIPS and Bayh-Dole undermined Brazil's and India's newly acquired economic and technological capacity by enclosing and controlling information. Advocates of policies like TRIPS and Bayh-Dole claim that resource (information) "waste" could be countered by rendering nature and life systems productive.[94] In controlling economies of waste and productivity, these governing regulations functioned to route and control information according to a model of development built upon debt relations and dispossession echoing the Bretton-Woods system so crucial to postwar American economic expansion. Revived in the dematerialized terms of the information economy, they do so in less visible ways through the harmonization of IPR standards. "Common problems/common needs," "shared risk"/"mutual responsibility": to whatever degree the language of harmonization advances warm claims of microbial kinship, commonality, sharing, and mutuality, it does so *in order* to open up the biosphere as a global commons from which materials can be taken. In this context, Michael Flitner has argued that, in contrast to the understanding of commonality as belonging to all and ownable by no one, regulations such as TRIPS and Bayh-Dole function to redefine "commonality as belonging to no one and thus ownable by some."[95] The language of harmonization works within this same interpretation and repositions developing countries as raw resources of bioinformation for biocapital production.

In Novo/Ex Nihilo

The language of harmonization works to keep the frontiers of the innovation economy open and endless. When *America's Vital Interest* insists on placing "innovation" at the center of its strategy against emerging microbes, we glimpse the multiple layers of waging war against microbial emergence: since emergence is endless, the report concludes, the best strategy is to siphon its novel creativity into capital by enclosing its mutant horizons within intellectual property regulations. Innovation once referred to the material instantiation of knowledge within a tangible object. Thomas Edison's lightbulb or Alexander Graham Bell's telephone were both patentable as material inventions developed from the innovators' knowledge. However, the meaning of invention and "innovation" changed along with the dematerialized terms of the information economy to include forms of "disembodied knowledge" alongside the "specificity" of its forms in intangible realms such as methods, plans, properties, and data.[96] The dematerialization of innovation renders highly elastic the place and moment of novelty. At what point is knowledge "new" once it is no longer required to become incarnate as a material thing? Along the same lines, what is the "novo" of innovation in the new information economy? The history of TRIPS and Bayh-Dole demonstrates that the global circulation, analysis, and usage of bioinformation operates as a way of "manufacturing innovation." By carefully orchestrating a dynamic of flow in relation to one of fix, this regulatory infrastructure controls the movement of information in ways that help to refresh a company's revenues (in the words of the IOM), thereby "ensuring [it has] the right amount of product in the pipeline" at any given time.[97] During the 1990s, American lobbyists and advocates for patent rights claimed that a failure to erect and enforce intellectual property laws would "extinguish the nation's creative potential" and block the innovation pipeline.[98] In controlling the spaces of informational transmission through intellectual property laws, the United States could maintain prices at artificially high levels in its primary markets, while securing its position as the leading maker and distributor of biotechnologies for developing countries. By directing the global flow of information, developed nations—primarily the United States, Europe, and Japan—would reap enormous economic benefits from the worldwide traffic in bioinformation.

The effects of this system are exemplified in the case of the AIDS crisis in sub-Saharan Africa. By 1998, South Africa alone had the largest number of people infected with AIDS: 83 percent of the world's new cases in that year.[99] At the same time, antiretrovirals were being broadly adopted and used in the developed world, transforming a fatal infection into a manageable chronic disease. Yet most people in sub-Saharan Africa could not access such extremely costly treatments. In this context, President Mandela hoped to invoke an emergency clause in the TRIPS agreement; article 31 of TRIPS formally permitted the use of patented pharmaceuticals without the authorization of the rights "in the case of national emergency or other circumstances of extreme emergency."[100] Yet throughout the 1990s, the United States made every attempt to discourage the use of this emergency measure. As a result, Mandela signed the South African Medicines Act into law in 1997, a step that enabled the nation to obtain critical pharmaceuticals through parallel importing and compulsory licensing.[101] India—at the time the largest manufacturer of generic pharmaceuticals[102]—became the country's primary supplier of generic antiretrovirals for as little as 1 percent of the cost of the American versions. In 1998, the United States retaliated against this measure by working through its trade and commerce departments and through members of the European Commission and the World Trade Organization (WTO) to initiate TRIPS violation proceedings against South Africa. Mandela was named as first defendant.[103] The American claim was not only that South Africa had transgressed WTO accords by depriving patent holders of opportunities in emerging markets, but that they had paved the way for other developing countries to do the same.[104] Controlling information worked as a means to locate and secure the "market share" of the global need for new biotechnologies.

"Innovation" was redefined as something that can come into being only once structured by privatization and property. Further demonstrating the resolving power of the marketplace, such regulations also posited that there could be no innovation without enclosing information in intellectual patents. Vandana Shiva, Michael Flitner, and Bronwyn Parry, among others, argue that while these regulations claim to advance invention, knowledge, and discovery, they have actually impeded all of these elements. Shiva asserts that, by using a Western metric of creativity, these regulations by definition side with the world's most powerful corporations. By standardizing an American-based interpretation of innovation, these IPRs make monocultures out of the rich and diverse knowledge emerging from the intellectual commons—farmers, villagers, indigenous tribes, forest dwellers, universities, and scientists alike. And although the knowledge and inventions claimed under IPR often derive directly from those intellectual commons, they are considered "innovations" only if they can be brought into being for industrial and commercial ends.[105] As Cooper wryly notes, intellectual property measures did not really function to protect the innovation economy. Instead, they "created it ex nihilo."[106] The regulatory infrastructure of the bioeconomy manufactured the innovation economy not by rendering possible the protection of new inventions or new knowledge, but by walling in and sequestering information. This control by limitation and enclosure enabled the production of monetary value from specific entities—whole specimens, biomatter, populations, and processes such as IP law, standardization, and financialization—where none had previously existed.

"Innovation" here is hardly synonymous with "invention"; it is the condition of being situated where the property of transformation itself becomes securely enclosed inside the sheltering fence of intellectual property laws. Operating through the language of harmonization, the United States sought to position itself at the forefront of global efforts to plumb the incipient frontiers of emergent microbes. The corresponding system of regulations, norms, and accords—from raising capacity to biodiversity discourse to TRIPS—represents a broad-spectrum effort to transform the world into a regulatory environment that generates economic productivity from life. It does so by channeling, accelerating, and protecting the movement of bioinformation across the globe in ways that endow it with economic significance and ensure regulatory compliance. When the WTO came into being on January 1, 1995, a year after TRIPS was ratified, it formed an international, intergovernmental organization that ensured member state compliance with GATT/TRIPS. Nation-states can be penalized for noncompliance, at least in principle. In practice, however, while the European Union and the United States most frequently violate TRIPS accords, more than 80 percent of the cases against them have resulted in no charges whatsoever.[107] Through this biased and uneven space of tolerance, the WTO has become a crucial mechanism in upholding the interests of information-based corporations in wealthy nations, above all the United States, whose demands developing nations are obligated to recognize, adopt, and enforce. Raising capacity is thus a U.S.-led effort to render the world compatible with the fluid economies of biosecurity. Crucially, the control of these economies manages informational flow in ways that harmonize the discordant world with the interests of the developed world.

In celebrating information as a medium of maximal flow, the fluid economies of biosecurity advance a form of global health seemingly unencumbered by the concerns of the material world. Dissolved into informational registers, bodies and networks, socioeconomic inequalities, colonial logics, and structural violences are transfigured into data points, weighted bits, and fractionated pixels. This conceptualization of information stems from the visions of postwar information theorists. These theorists saw the promise of a medium that would allow them to strip away the material complications obscuring the pure patterns of data that they believed made up the world. As I suggested above, these historical contingencies deeply shaped the forces through which information became the medium of a new paradigm of development. That the world could be distilled into informational patterns meant that it would be possible to access, understand, and manipulate that realm through the powers of information. This was possible because, in its ability to break apart, neutralize, and meld opposing forces, information offered an ideal medium through which to address as one and the same the divergent aims of preserving life and creating markets (the two realms made to dovetail in the term "bioeconomy").

These fantasies of information form the basis for the illusion that a vastly uneven and unequal world could be changed without confronting the difficulties of geography, politics, and conflict. The production of information along these transcendent lines helped to shape the commonly held conviction that global health is best managed as a data-driven affair. In apprehending and managing global health as a data-driven matter, global health is depoliticized. It ceases to entail redressing inequalities, providing clean water, improving nutrition, or addressing climate destruction—the very factors that the IOM names in *Emerging Infections* as the cause of disease emergence.[108] Instead, it responds to the imperatives of "resilience." Global health can be obtained not by addressing the actual structural health needs of the world, but by making communication more efficient. Emerging from Cold War informational fantasies of apprehending everything from life systems to social systems (and, as we shall see, economic systems), the informational strategies deployed by Wolfe's GVFI regard the computational as the ideal response to all social ills.[109] Recall that he first repurposed his AIDS research lab in Cameroon to make way for the first GVFI laboratory. This is a salient reminder of the dynamics of this larger system (of which Wolfe's work is only one contemporary evolution); it ignores existing needs in developing countries and seeks technological expansion for developed worlds. No longer, then, is the emphasis on preventing catastrophic microbial futures, but on rendering always at-risk futures resilient and responsive to their unforeseeable movements. This effort is not premised on making use of bioinformation to address present needs, but on crafting global conduits to structure and protect its inestimable future value. According to this vision, individual nations must be made into interoperable components in a larger technological system, rather than being made healthier or more stable. This version of global health no longer treats inequality as a political problem: it is now a puzzle of technology. The world can be designed out of crisis simply by growing technological capability in developed nations, enabling their innovations to keep pace with the creativity of emerging microbes, and by perpetually honing techniques that keep their fingers on the pulse of its constant change.

The grey literature examined in this chapter reveals the conduits and regulatory infrastructures through which the vagaries of microbial emergence can be produced as endless bioinformational resource. The fashioning and use of these currents and networks in the service of the now-global war against emerging microbes is what I have been calling the *fluid economies of biosecurity*. Paying attention to these flows reveals a system that maintains and defers insecurity and security under the promise of development and equality. Though premised on commonality, synchronicity, and commensurability, this effort nonetheless relies on internal inconsistencies, inequality, and nonsynchronous displacements for its optimal functioning. What becomes apparent when we examine contemporary biosecurity in light of its fluid economies, then, is the production of insecurity not as a paradoxical opposite of biosecurity, but as consistent with it. With these contradictions in view, I traced the reverberations of the harmonization project out into the world with a focus on the precise ways in which the innovation economy works in global health to enclose ecologies, politics, and social relations within its crisis-perpetuating logic.

The ways that innovation and perpetuated crisis management

become interlocked in this process demand analysis. How does it matter that this dynamic suffuses a global project whose administrative organization entails the reshaping of space and time for its smooth functioning? The answer depends on understanding the bioeconomy as a modern logistical system that works by mediating between the intangible realms and processes of virtualization and the transport, storage, and transmission of material goods. I take up this task in the following chapter.

5 Managing the Microbial Frontier

This chapter examines why the project to secure emerging microbes repeatedly fails, even after four decades of concerted efforts around global health and biopreparedness. I answer this question by returning to the same object and archive considered in the last chapter: the grey literature around the language of harmonization, and Global Viral Forecasting Initiative's (GVFI) dream of creating a planetary immune system. Here I shift analytical registers to ask after the administrative organization of that global project and its irresolvable contradictions. Although often associated with cosmopolitan concerns over human welfare, the project of global health is closely implicated in post–Cold War U.S. economic and security concerns as they interact with the disparities of global capitalism. I bring these dimensions of the bioeconomy into view by reading it not only as a process made up out of the realms of political discourse, but also as a built system. I argue that the making of the bioeconomy is a problematic of logistics: a puzzle of reshaping the world into a technological system that enables the movement of bioinformational goods while controlling and maintaining conduits for their circulation in ways that generate and sustain capitalist value.

To begin this reframing of the bioeconomy, I return to GVFI's website logo from another angle (Figure 4.1). Through its image of earth floating within a polyhedron, our perspective is aligned with the omniscient viewer of IBM's *Data Anthem* advertisement (chapter 3) who, in scanning the world as data, sees everything from all perspectives at all times. Because this gaze is without a body (and therefore without a subject position), it is equated with the production of universalizing knowledge of the world.[1] This privileged perspective claims to see into the world's hidden order, implied in this graphic by geometry; the polyhedron promises to disclose the concealed rhythms and dynamics that govern and organize this world. The geometric lines crisscrossing and encircling the world suggest networks through which all points on the globe are interconnected. These lines are also conduits of circulation, simultaneously and always connected to all other lines. The view of a globe displaces the axial point from which those in the Western world are accustomed to seeing it; instead of a view that privileges the centrality of North America and demarcated nation-states, here the lines converging in the North Atlantic connect that region to all the Americas, Europe, the African continent, and the rest of the world. Landmasses are not marked by borders, but are rendered in the sameness of a flat and even grey. GVFI's logo could readily be used by a global logistics company. A Google search for "global logistics" will likely pull up reams of stock images that share these themes: the world as uniform surface upon which things can flow, governed by moving systems whose global connections remake it as a distributed network. Under logistics, the universal perspective of God is replaced by the infinite managerial vision of always-on twenty-first-century global computational systems. I make this comparison not to suggest that the bioeconomy is *like* logistics. Instead, it *is* a logistical system.

In the previous chapter, I examined how the biosecurity project called up by the war against microbes reshaped the world into a vast platform for informational generation, processing, and transmission to make up and maintain the currents and currencies of the bioeconomy—a process I call the "fluid economies of biosecurity." These economies pursue an immaterial horizon: innovation. They traffic in ostensibly intangible things: information. They seek to capture a nonobject: potential value. They are shadowed by juridical documents that record political and economic relations, but whose content can never quite achieve the status of an analyzable object, image, or representation. For all its disincarnations, however, the bioeconomy is a global commodity chain at its core.[2] As a commodity chain, it inhabits and spans geographies; it flows through a series of interconnectable, interoperable components and conduits designed for its support; it binds actors, infrastructures, and institutions together around its affordances. Like more overtly physical logistical systems, the seemingly immaterial realm of invisible processes that constitute the global bioeconomy are made up of rules that condition its functionality, legislative agreements that dictate its transfer points, policies that plot out its distribution networks, resolutions that embody the political rationalities underwriting its technological forms. For these reasons, the bioeconomy must be read against exclusively dematerializing trajectories and subjected instead to a line of questioning more capable of tracing it out as a material system, of which the grey literature of regulation is itself a crucial component. The previous chapter showed how the production of new bioinformational goods met the changing needs and imperatives of the emerging postindustrial manufacturing economy. Global space was reformatted through regulation into a vast platform for informational generation, management, and transmission. Bioinformation may be the content of this commodity chain, but its very existence relies on a body of regulation that transmutes the world into an informational orb. In order to bring this infrastructural substrate into focus, I move the commodity to the background.

My methodological approach in this chapter enacts this reversal of foreground and background. I read the regulation around harmonization as global infrastructure through which the bioeconomy, and the bioinformational commodity itself, could come to exist. In this sense, infrastructure may be thought of as "objects that create the grounds upon which other objects operate."[3] From this perspective, financial mechanisms, management strategies, and administrative techniques appear as infrastructural forms in their own right. While infrastructure studies necessarily emphasizes materialist analyses like bridges and sewer systems, regulations too function like designed networks that generate environments, and are "things that are also the relation between things."[4] The grey literature reveals the mechanisms through which the world is rewritten as a series of conduits, codes, transfer points, and networks that determine how goods flow. Just as sewers diverting water constitute a system of substrates that permits the world to operate in a relatively seamless way, economic logics, administrative processes, political rationalities, and managerial techniques form the substrate of the bioeconomy. In turn, the bioeconomy is itself the material evidence and manifestation of its specific infrastructure. This treatment of regulation as infrastructure reveals the project to raise capacity as a logistical enterprise that reformats global space and time in accordance with the requirements of global capital.

Logistics is the managerial art of distribution, the processes through which raw resources are transmuted into consumable commodities, the military science of fulfillment, the transposition of military systems into the management of everyday life, the channeling of flows to move and maintain supplies across space and time.[5] In each of these definitions, logistics appears as the art and strategy of mediating between the abstract and concrete to engage with a world of flows. Logistics volleys between immaterial and material registers as it forms a managerial system that shapes the world temporally, spatially, and geoeconomically. In returning to the language of harmonization and the U.S. project to raise the world's capacity for global health, I examine how the bioeconomy operates within this technical system that reformats space/time as a medium capable of routing goods to keep time with a new worldwide project. The project to raise capacity thus effects an "infrastructural globalism"—a durable and structured global project to produce data about the world through a shared technical base.[6] The previous chapter demonstrated how harmonization sought to draw different state interests, metrics of value, rhythms and cycles of production in line with the needs of a new economy in bioinformation. In this chapter, I question the extent to which these diverse tempos and processes of production, and divergent goals and metrics of value, are homogenized in this effort to make the world keep pace with the newly standardized norms. Tracing the bioeconomy as logistical system makes visible the ways in which it effectively loops the world together in line with a normative rhythm of imperial measure. To analyze the bioeconomy as logistics is to see past the transcendent claims of GVFI's speculative and data-driven global pandemic surveillance project. This project spans the unending spatial and temporal frontiers of emergence even as they are staked out and claimed in advance. My reading sheds light on the neoliberal promise that animates broader global health paradigms, thereby revealing the inequalities written into its architecture. The bloated footprint of the now-global fight against microbial emergence is a crucial, if often illegible, expression of twenty-first-century U.S. expansionism.

One meaning of the term *resolution* points to the structural dissonance inherent in the creation of harmony. Holding that definition in mind, this chapter opens up a more complex dynamic involved in the project to harmonize. While it is often held that logistics aims for unimpeded flow, I analyze the bioeconomy to show how its logistical system carefully counterpoints fix and flow, lubrication and friction, connection and interruption, fulfillment and withholding. I explore some of the implications of building a global health project through such dynamics. I draw attention to the ways U.S. government bodies work within the logistical architecture of the bioeconomy through regulatory mechanisms that, in turn, enable certain countries to fashion novel forms of value out of the endless frontiers of emerging microbes. Whereas chapter 4 showed how the neoliberal drive melds a series of contradictions by collapsing values and interests of global health and militarization, commonality and privatization, biospheric life and commercial markets, chapter 5 shows that, rather than a seamless merging, this approach establishes a series of irresolvable tensions in the now globalized fight against emerging microbes. I read the grey literature to show how the United States works in these spaces of contradiction to craft synergy between crisis and value, emerging microbes and emerging markets, the endless horizons of emergent risk and "frontier markets" (the making of novel goods for new consumers in the neglected markets of less-developed countries). I track a global health project premised not on the goal of stopping or significantly slowing the process of microbial emergence, but on managing microbial frontiers *in order to* arouse more productive resolutions from their risks. This is one of the most potent effects of the concept of emergence: the novelty of emerging microbes—their potential to appear anywhere, at any time, in unending new genetic combinations—promises boundless economic production founded on the possibility of perpetually infected futures.

The Logistics of Harmonization and the Making of New Global Standards

To begin with a definition, logistics encompasses the entire life cycle of a product, from raw resource to consumption.[7] IKEA, for example, identifies production costs for each item it sells and plans the logistics around it. It extracts lumber from a forest in Eastern Europe. Other materials are sourced from suppliers in places like Pakistan. Those resources are then transported to factories in China where they are manufactured into a range of products. These products are packed into boxes, and each box is labeled with a barcode tracked via a vast network of computers as it moves from factory conveyor belts to cargo carriers to large shipping containers for distribution around the world. Some products move by railways that stretch across continents, while others cross oceans from ports such as Shenzen, China, to Long Beach, California. They are unloaded into containers, moved to holding centers, and finally to stores for purchase. Beyond this specific example, logistics is a system in which all manner of objects—washing machines, flip-flops, coffee beans, ball bearings, frozen shrimp—are interwoven through technological networks (e.g., shipping manifests updated in data centers), the built environment (e.g., from the geography of raw resources to store parking lots), systems (whether the vagaries of weather or war), and infrastructures (from ship-to-shore cranes to highways and beyond). Because efficiency of movement is inextricably bound to cost and profit, logistics aspires to flatten, lubricate, and connect the world. But the world is made up of disparate, disconnected, and unrelated components, entities, and processes; logistics must transform this into a vast space of interplay in which oceans and computers, highway bridges and transport trucks, landscapes and weather systems become interoperable. "Logistification," what Jesse LeCavalier describes as the historical and industrial process and practice of world making for logistics, organizes the world's gravity, drag, interruption, disconnection, and friction into an even surface of fulfillment that maximizes profit.

NATO defines logistics as the work of moving "the right thing, at the right time, to the right place."[8] This formulation captures the inner heart of logistics: the aspiration to resolve, overcome, and ultimately shed the friction that arises at the intersection of matter, time, and space. Logistics seeks to render the world as a smooth platform for the "just-in-time" domain that is its promise. This "just-in-time" logic underpins an inventory management system that controls the pace at which raw materials, resources, and products are released and received in the supply chain. It works to manage the temporalities of goods moving through the world. By dispatching resources and goods only as they are needed rather than days or weeks before, and ensuring that the right things arrive at the right time to the right place, "just-in-time" logistics avoids the financial losses incurred from unnecessary inventory holding costs in storage and increases turnover. Yet just-in-time logistics reveals that, in contrast to the logistical dream of pure flow, actual operations require stagnation, deceleration, and obstinacy. While those elements are often understood in the critical study of logistics as the result of accidents, unforeseen interruptions, and infrastructural breakdown, they are integral to its system of value.[9] Logistics needs friction in order to function. As a logistical system, the bioeconomy operates through strategically controlled slowdowns choreographed in counterpoint to free and fluid circulation. This control and coordination fashions the bioeconomy along distinct rhythms of production and consumption, waste and generation. I examine this dynamic of control, friction, and fix to show how a global health project built on this system ensures that the war on emerging microbes was a profitable economic venture for the most powerful nations.

If the logistics of the bioeconomy renders the world a smooth space of interplay, the U.S. project of harmonization confronts a corresponding problem: how can a socially and economically uneven world become an even platform capable of supporting the bioeconomy? From the standpoint of those invested in this conjoined project, poorer nations could better their economic prospects by becoming directly connected to a global informational economy.[10] From this perspective, information must glide smoothly through the world; global space simply needs to be reshaped to enable its flow, which (if properly managed) will enable an economically inequitable world to find its own level. To offer a sense of this logistical challenge, I return to GVFI's vision of a planetary immune system. Wolfe describes his work with scientists, technology specialists, and logisticians to fashion the world into a system in which biological samples could be transformed into glittering streams of data in the Silicon Valley control room, to serve the interests of what has now become a project combining the biological, militaristic, and economic. To transport biomatter from its "far flung viral listening posts" to its eventual destination within the ever-expanding spaces of speculative security and economics, this system choreographs disparate parts: squads of collectors (the bushmeat hunters), automobiles and airplanes that transport samples through biological processing centers around the world, diverse technologies with their own material and technical demands of storage and transmission. Its workings are articulated across different geographies, bodies, buildings, laboratories, systems, computers, devices, and servers. Wolfe's "global nervous system" for pandemic forecasting necessitates a technical approach that can pull disparate landscapes of extraction, production, storage, and transportation into a single structure. Oriented around the shared project of generating bioinformational bounty and sustaining its flow, GVFI's planetary immune system requires the disconnections of global space to be transformed into a smooth logistical continuum of pushing and pulling between extraction, production, storage, and transmission.[11]

Like GVFI, the U.S. project of harmonization seeks to rewrite the world into a universal format, a vast operating system capable of moving bioinformation through space and time. It does so through regulation and standardization. Standardization enables logistics to make the world an interconnectable system, eliminating all variations by introducing common systems of measure, valuation, and validation. It rewrites and reconfigures bumps into a smooth terrain of formulaic sizes and shapes, and harmonizes formerly disparate parts within a system of shared measure. Thanks to standardization, a logistics planner can trust that the pitches of a screw manufactured in Milwaukee will ease into the threads of auto parts made in Guangzhou. It ensures predictable conditions on every scale across the globe, from the cardboard box containing the screw to its placement into other boxes, onto boats, across oceans, into distribution centers. Standardization determines the material specifications of the screw, the measurements of the crane braces that grasp and lift the containers into which it is eventually packed,[12] the twist-locks that fasten this freight onto transport trucks, and the highway bridge allowances designed to clear a container's maximum standard height of nine feet, six inches as it moves cross-country. Standardization renders a world of quirky and incompatible variations contiguous through specific arithmetic guidelines. It functions as a globally shared language of calibrations, dimensions, calculations, and measurements that enables the world to become a flowing space of interplay. Without it, nothing in logistics would be able to communicate.

Global standards are the lingua franca of global circulation. Adhering to these rules is a precondition for inclusion in systems of worldwide exchange. For a given actor—an institution, a firm, a laboratory, a nation-state—to participate in its networks and benefit from the promises of global interconnection, they must first align with the standard means and measures dictated by universal infrastructures. The process of standardization may seem a dry, technical, apolitical process through which standard-setting organizations work to craft happy mediums that facilitate the world of connectivity. However, standards arise from deeply political and often contested processes, shaped and initiated by state actors that represent the interests of corporations or government bodies.[13] Nation-states lobby to secure favorable global market conditions for these entities by writing their technical specifications into international standards.[14] Behind a seemingly neutral universalism lies a charged struggle over who will dictate the terms through which things connect and move through the world. In this sense, standard setting is a product of political and economic power relations. International standards thus reflect the preferences of dominant countries, and the countries that shape the majority of international standards—the United States and Western Europe—become the architects of the infrastructures of global distribution.[15] Keller Easterling calls standards a "'soft law' of global exchange":[16] a technopolitical strategy to lay out and shape the corridors and conduits of global connection, and to set down the conditions that others must meet if they are to operate in and benefit from these systems. Infrastructure possesses an agential power to determine what is possible and what is not: the international standards through which it is built dictate its parameters. The ability and willingness to harmonize with these standards determine the extent to which a given state, actor, or set of interests can gain access to a world of economic and political opportunities.

The U.S. effort to harmonize is imbricated in the same standardizing logic. Its mandate to raise the world's capacity for global health calls upon developed nations to assist developing ones to enter the fold of new standards. GVFI's planetary immune system again offers a valuable example of the dynamics of raising capacity. Wolfe describes how GVFI works within its field sites to incorporate so-called viral hot-spots into a global health surveillance network. "Working with" involves intangible elements like "educating bushmeat hunters" and wet-market merchants about the dangers of hunting, training them in specimen collection, teaching them and their fellow villagers about "proper" sanitation and hygiene habits, and requesting that they monitor and disclose their illnesses and report evidence of animal die-offs in the deep forests. It also includes material processes and practices such as building collection and storage facilities and distribution networks that can move a virus from the field to a lab, emplacing lab protocols, and encouraging the purchase and use of new technologies (such as cell phones) to enable "risky" people to submit reports while being geotracked.[17] These measures embed sites and systems of knowledge, ethical perspectives, and, more pragmatically, relevant technologies and equipment across the globe. Later I show how this effort was braided together with opening new markets for biomedical technologies and products from the developed world.

The project to raise the world's capacity works at the level of standard setting to influence global exchange in ways that structure how bioinformation can move. Just as standardization in logistics compresses the heterogeneity of the world in the service of flow, the harmonizing effects of raising capacity seeks to level a mottled world, making it into a common format where bioinformation can, at least theoretically, move seamlessly anywhere. It approaches the challenge of a socioeconomically and technologically uneven world by seeking to transform the global regulatory environment in ways that make it possible for those underdeveloped countries and their "risky" populations to keep step with newly standardized global discourses, measures, goals, and requirements. Raising capacity through harmonization can thus be read as part of a larger U.S. project to achieve what Paul Edwards terms "infrastructural globalism." Unlike "voluntary internationalism"—in which states share data around a collective object (for example, weather systems) out of a temporary confluence of interests—infrastructural globalism describes projects that build long-term, world-spanning networks rooted in global epistemic communities and technical bases, in which sociotechnical systems interconnect the earth as a globalist site of shared action, knowledge, and governance.[18] The effort to raise the world's capacity for global health surveillance seeks to emplace durable ethical, legal, scientific, and metric global norms for collecting, storing, processing, moving, interpreting, and valorizing data. It enacts a "a quasi-obligatory globalism based on a more permanently shared infrastructure."[19] To make, connect, and expand networks for global pandemic surveillance, these new norms of global health must take hold across the world.

The Circuits of Global Health and Normative Rhythms of Imperial Time

The regulatory infrastructure through which norms of global health are fleshed out expresses otherwise undeclared political interests under the presumption of universal interest. Yet regulatory infrastructures do not enable individual nation-states to act in totalizing ways, or to shape a world around a unitary vision. On the contrary, infrastructural globalism is never fully achieved. The project is inevitably riddled with incoherence and confronted with both political and material difficulties; it never operates as imagined. But however piecemeal and patchy such efforts appear to be on the ground, their infrastructural effects advance the power of specific nation-states, actors, and corporate interests above others. Regulatory infrastructures arise in interaction with complex global formations—the overlapping sovereignties, for example, of international organizations like the WTO, IMF, and WHO.

Nations seeking to become a part of these systems must make a series of material, symbolic, political, and financial commitments that empower infrastructural globalism. Even as regulation works in complex spaces beyond state jurisdiction, it operates at the level of standards to further the interests of certain actors while pressuring others to fall into step. The call to harmonization thus encourages, prods, and enables poorer nations to meet globalized moral, scientific, technological, and administrative norms in the global health project. Because these normative measures and values are heavily determined by wealthy Western countries, the movement, ownership, and value of bioinformation are marked by unequal power relations.

The project of harmonization thus replicates and deepens long-standing colonial practices and logics. Although raising capacity was put forth as a new paradigm of development, the dematerialized and democratizing language of the informational economy harbors systemic violences that are difficult to discern. Tracing out the regulatory infrastructure that brings this project into being can draw this neocolonial dynamic into resolution. When tracked infrastructurally, regulation appears analogous to undersea telegraph lines, submarine internet cables, broadband towers, and other worldwide systems whose material nodes and networks (the environments they occupy, the populations and interests they connect, the histories and politics they instantiate) are rendered possible because they trace over existing imperial dynamics, even as the latter are changed by ongoing political and cultural negotiation.[20] The project to raise capacity is designed to locate and channel resources, to categorize goods, and to divert value. Like other global infrastructural systems, those involved in harmonization trace over and reinforce colonial legacies of resource piracy, appropriation, and dispossession in the informational realm.

While colonial geographies of collecting, storing, and usage are often analyzed as a geographic problem, Parry argues that the bioinformational economy constitutes a different kind of space: one that re-places colonial dynamics within the virtualized realm of the informational. In previous models of colonial collecting, specimens were acquired, ordered, controlled, and recirculated across different social sites, and this movement is a value-adding mechanism.[21] Informational techniques and technologies such as genetic coding, high-throughput sequencing, and real-time databases have flattened these social spaces of exchange. This launders the spaces of colonial appropriation into a generic one of "global exchange." At the same time, the geographies of colonialism become condensed into a singular space/time of the informational. This shift from the geographical to the informational accelerates the processes by which resources accrue value because they enable collectors (most often those who control the economic, technological, and regulatory conditions of acquisition) to route and control them in a now-compressed space—what Parry calls the space/time of "speedup"—in which power, profit, and collecting are now collapsed.

I argue that this space/time of speedup administers the world of plural temporalities and heterogeneous values in ways that make it match up with a "universal" measure. Whereas many in non-Western worlds might define terms like "innovation" as the social processes that give rise to diverse forms of knowledge, "wealth" as having enough to support the needs of the earth and its inhabitants (and, often, of having enough to share freely), and "generativity" as the force in life systems that enables their renewal and change, the project of harmonization overwrites plural rhythms of productivity, definitions of value, and paces of transformation.[22] It does so by redefining values in ways that matter only to neoliberal globalization. By rendering diverse metrics and heterogeneous temporalities into the common format of information, and by rerouting them through the space/time of speedup, regulation works as the content manager of this informational realm; it channels bioinformation in ways that obligate all to answer to the demands of the bioeconomy through a homogenizing process. It binds the temporalities, processes, and objectives of various world systems, in the broadest sense, to the accelerated speculative spaces of twenty-first-century economic ventures and security paradigms. By making these heterogenous values and tempos fall into step with the bioeconomy, these variations are brought into a single harmonious time set by developed nations. This normalizing pace is dictated by the appetites of the innovation economy: it makes different systems and processes serve the economy's demand for limitless productivity. The strategy of harmonization, then, seeks to loop a heterogenous world into the normative rhythms of imperial measure.

Yet the regulatory infrastructure that makes and governs the logistical workings of the bioeconomy did not work to protect the property of novelty as such. Instead, it mobilized two countervailing forces: one to move and free information, the other to restrict it and so control the pace of innovation. This logistical system generates information and creates flow on the one hand, while creating choke points and friction on the other; it takes advantage of the informational commodity in its capacity to become more valuable as it is recombined, replicated, or transformed, through expanding its spaces of transmission in ways that add value.[23] In this vein, several scholars argue that in supply-chain logistics, power lies not only in the ability to speed circulation up, but also in retaining an active power to slow it down.[24] The analytical literature often examines logistics primarily as an effort to move more goods with greater speed through supply chains. Because the accumulative process can be intensified by "annihilating space by time," this literature emphasizes the way that logistics seeks to obliterate spatial barriers, compressing them via time.[25] Yet the lulls and surges of just-in-time logistics reveal a more complex, "pluritemporal" dynamic at its core.[26] It produces and increases value for the bioeconomy by arranging speed and flow in specific relation to points of slowdown and blockage.

Having examined how harmonization seeks to administer the bioeconomy along a normative temporality of imperial logic, I now consider how regulation is infrastructural to shaping its spaces of transmission in ways that determine who profits most from this project. This sets the stage, in turn, for analyzing how that neocolonial logic is diffracted across the spatiotemporal frontiers of microbial emergence.

"Doing Well by Doing Good" in the Emerging Marketplace of Global Health

As I argued earlier, U.S. economists saw the new project of global health as rich ground for a robust innovation economy. The 1997

National Institutes of Medicine (IOM) report spells out the reasons for the U.S. investment in global health: *America's Vital Interest in Global Health: Protecting Our People, Enhancing Our Economy, and Advancing Our International Interests.* Through its very title, the report frames emerging microbes as the promising domain through which the United States can create an expansive "global health-related economy."[27] This new marketplace summons all industries engaging in bioinformation, as well as those invested in the mutable and circulatable properties of informational goods as a whole: all the industries that build, provide, operate, and use the goods, services, and information available over its ever-evolving networks. It encompasses the world of devices, knowledge, pharmaceuticals, vaccines, and software. It also requires the cooperation of all countries—bilaterally, regionally, and through multilateral organizations—to connect their respective networks and share information. *America's Vital Interest* advocates for U.S. involvement in global health as a way of developing American science and technology across this rich terrain.[28] Emerging microbes function as an immaterial and therefore inexhaustible proving ground for the development of all global health-related technologies.

Just as the policies of the New Right conjoined the economic and the biospheric (chapter 4), this neoliberal spirit was at work as this marketplace enfolded once-separate realms into one another. Although often associated with cosmopolitan concerns over human welfare, the concept of global health is closely implicated in the neoliberal economic development of the United States, which seeks to reach humanitarian goals by enabling the free market to move unfettered in its domain. Indeed, from 1990 onward, the World Bank began to replace the WHO as the central international body in charge of directing, financing, and administering the global health project.[29] It increased its loan financing for global health from $317 million to $2.3 billion annually, and then, after the Agreement on Trade-Related Aspects of Intellectual Property Rights (TRIPS) was ratified, to $13.5 billion. The World Bank became "the single largest external source of global health funding for low- and middle-income countries."[30] *America's Vital Interest* deploys this disparity in power and economic clout to cast doubt on the WHO's ability to mobilize resources and perform its normative function of setting international health standards. The report suggests that the humanitarian goals of global health are best met by the IMF and World Bank and The General Agreement on Tariffs and Trade/World Trade Organization (GATT/WTO).[31] The growing influence of these economic institutions in global health signaled its recasting as an issue of financialization and trade.

In this reconstitution of the global health-related economy, "the marketplace" was seen as capable of melding public (human) welfare with corporate demands. The goals and principles associated with more humanitarian paradigms of global health were not so much reversed (as some scholars suggest) as they were differently mobilized and rerouted to dovetail with the pressures of neoliberal capitalism.[32] Global health—now conceptually and institutionally conflated with the marketplace—was reshaped to serve a dual purpose. Although such a project could only ever be replete with divided interests, divergent perspectives, and conflicting priorities, "the marketplace" was produced as the shared ground in which all constituencies could be invested. As a consequence, the United States could confidently engage with the exigent need for health in that marketplace. The authors of *Orphans and Incentives* argue that the often contradictory interests and different priorities of various actors would be reconciled in this newly refashioned domain. U.S. government reports similarly describe "the marketplace" as the common terrain where the diverse aims of a broad range of actors throughout the globe—medical professionals, public researchers, technology and pharmaceutical firms, investors, populations in need of better health, the United States and other nations, and more—would "ultimately coincide."[33] According to this logic, the needs of a Congolese bushmeat hunter and a data analytics startup firm in Palo Alto meet in a rosy "win/win situation," the ultimate resolution of conflicting interests. "The marketplace" was viewed as a leveling medium that purportedly massaged out inconsistencies and mismatched interests, those recalcitrant bumps and points of discord on the road to harmonization.

Needless to say, those same reports make it abundantly clear that the United States would be the rightful and primary beneficiary of the new global system. As *Orphans and Incentives* asserts, the United States must meet the global need to address emerging microbes from a position of "enlightened self-interest."[34] Consequently, *America's Vital Interest* argues that the United States must articulate its involvement in the project of global health in order to "do well by doing good."[35] The report refers to UNICEF's children's vaccine program as an instructive case. Under this global program, European and American vaccine manufacturers tendered bids to UNICEF to supply vaccines at lower prices for use in low- and middle-income countries. Controversially, price-tiering enabled these corporations to sell vast quantities of pharmaceutical goods to lower-income countries for less money while maintaining inflated prices in domestic markets to cover the full cost of production, investment in research and development, marketing, and overhead.[36] The United States withdrew from the program in 1982 because many American politicians were against "subsidizing" the cost of vaccines for middle- and low-income countries, and American vaccine developers did not believe it economically worthwhile to engage with the critical needs of lower-income markets.[37] Yet when a 1994 review by Mercer Associates revealed that European manufacturers were profiting from these markets while making significant inroads in new venues for a range of biotechnological products, the United States perceived the need to reverse course.[38] Citing the decision to withdraw from participation, IOM argued that the United States had inadvertently "caused a loss in comparative advantage in this global marketplace" because it had cut itself off from access to such a high-yield market. In response to this apparent misstep, the IOM announced its commitment to reenter the program. Doing so would afford the United States rich opportunities to save children's lives by "sell[ing] large volumes of low-priced vaccines to the poorest countries of the world."[39]

For the IOM, the United States should address matters of human welfare not only because it enabled them to keep prices exorbitantly high in domestic markets, but also because meeting the health needs of less developed countries opened up a highly generative market.[40] This new market was valuable for two major reasons. First, many experts at the IOM perceived that the United States could make up in volume what it would sacrifice in price by engaging the needs of these newly emerging consumers. Second, these consumers enabled critical inroads into "frontier markets" that would help the United States gain the greater share within the global bioeconomy.[41] *America's Vital Interest* yokes U.S. economic expansion, foreign policy, and international development mandates to the broader conviction in neoliberal economics that all social needs are best met by the natural creativity of the free market. The report ties the success of global health to the precondition that it must be "economically feasible for the best American science, technology, and industry to address major global health problems *and* enable US industry to profit rather than suffer losses by that engagement."[42]

To foster involvement in these frontier markets of global health, *America's Vital Interest* states that the U.S. government, along with its counterparts throughout the world, should court the private sector to form new alliances. To accomplish this aim, the government had to ensure the regulatory, legislative, and market conditions necessary to attract private investment in information-related industries, from telecommunications to gene data banks, from information services to biosequencing labs. Bayh-Dole and TRIPS were cited as crucial to this process of linking state and corporations together, harmonizing their agendas to make the U.S. "a more enabling investment environment" for the private sector.[43] As I have shown, Bayh-Dole and TRIPS were two notable measures of patent policy through which the United States sought to control the process of change involving bioinformation.[44] Bayh-Dole gave rise to a system whose signatures are the joint venture, the public-private partnership, the venture capitalist, and the newly minted scientist-entrepreneur, many of whom (like Wolfe) initiated startups in Silicon Valley.[45] Those two pieces of regulation incentivized private corporations to take and use public research on emergent natural systems. Together, they formed key elements of the mechanism through which the properties of microbial emergence were framed as vital both to the global public good and to particular corporate interests.

This expansion of the U.S. bioeconomy was not limited to the war against microbial emergence. Emerging microbes was one significant and visible part of a larger effort to reach and retain "emerging consumers" for biotechnological and informational goods, enabling the world's poor to access a wide array of U.S. products. Framing health as an inalienable right *as well as* a critically unmet commodity, *Orphans and Incentives* recommends that the United States seek every opportunity to locate and engage "frontier markets" in global health.[46] Along these lines, *America's Vital Interest* suggests that "training health professionals [raising capacity] and researchers from the developing world would have other benefits as well. Health professionals who would return home with knowledge of state-of-sciences methods and medical technologies would be in a position to take scientific advances from the United States and make them relevant to their own countries. This would provide opportunities for U.S. medical products and technologies to enter the overseas markets."[47] Following this strategy, the United States would sell biomedical products to less developed countries as a "social good" to assist regions in need while actualizing significant financial value.[48] Michigan-based marketing guru C. K. Prahalad celebrates this strategy as the practice of finding "the fortune at the bottom of the pyramid,"[49] framing it as an effort to develop profitable "new markets around the needs and aspirations of the poor," those millions around the world who lack basic health necessities. The technique lauded by Prahalad is constructed around a new valuation of the poor by the private sector, arising from the division of the world into "bioinformational resources" and "information managers." The IOM identifies poorer countries as both the rightful recipients of humanitarian biomedicine and as "a market with [yet] unfulfilled potential."[50] In these terms, the report starkly clarifies the link between emerging markets and U.S. economic nation building, stating that the United States' "economic success is a precondition to global health." As a logical consequence, it continues, "we must pursue growth in emerging markets as aggressively as possible."[51]

The unevenness produced by such proprietary approaches emerges in the specific ways that elements of this global health model are defined. For the IOM, the "goods" to be generated through public-private partnerships are not those that address the structural causes of ill health, factors that would actually impact the pathways through which microbes change and develop pathogenicity, or sociotechnical or biomedical interventions like vaccines for existing illness. Instead, the IOM emphasizes products that can make the most of the ever-mutable and mobile qualities of information, focusing on speculative technologies like data modeling, prediction, computational biology, and data analytics that purportedly monitor the changing horizon of emergence.[52] Mirroring the logic of dissipative life systems, neoliberal discourse views itself as a system that generates innovation from crisis. If the United States can locate the "fortune at the bottom of the pyramid" by connecting with emerging markets, then profit inheres in the self-renewing structure of emerging viruses. As the authors of *America's Vital Interest* state, "it is the boundless potential of innovation, rather than price" that gives the United States its competitive advantage in the business of global health.[53] Once fabricated and enclosed within the new regulatory apparatus, "innovation itself"—this time in the form of a self-renewing demand for goods related to the perpetuated risk of microbial evolution—becomes the primary means through which corporations can retain "emerging consumers."

Emerging Consumers at Emergence's Unfolding Frontier

Standardization, harmonization, leveling, oneness, mutuality, sharing: these efforts and processes to level unevenness are a precondition and ostensible effect of global public health. But the language of harmonization produces a range of incompatibilities and tensions—evenness and differences, speed and drag, informational freedom and control. I turn now to examine how the United States worked within this space of contradiction to craft new forms of synergy between emerging microbes and emerging markets. Paradoxically, the evenness implied here is not regarded as a goal to be reached. The drive underpinning and moving this effort forward must refuse harmoniously leveled conditions as part of its internal logic, precisely because such smoothness and stability would eliminate the dynamic tension at the heart of market growth. Instead, the project of microbial resolution must work in the productive space that constitutes the gap within its own logical contradictions: while seeking to level uneven conditions, their total eradication would erase the driver of the project. In order to become a rich basis for the corporate exploitation of global health inequalities, emerging microbes must remain unresolved. Cooper illuminates this problematic: "Rather than seeking to obtain this goal, neoliberal capital confronts that goal as the limit to economic growth, and, therefore seeks to deflect it."[54] Here is the powerful and relentless out-of-sync dynamic that underlies the "just-in-time" promise of seamless delivery of effective health measures. The project of microbial resolution promises to mobilize information in ways that enable an uneven world to "find its own level." But the commensurability implied by each of these discourses is not posited as a concrete goal to be definitively reached. The drive underpinning and propelling the militarized quest for global public health must refuse, as part of its own internal logic, conditions of evenness. In contrast to the vow of fulfillment promoted in this vision, the logic of harmonization demands that life be devalued and set into continual circuits of crisis.[55] Only then can crisis be transmuted into an endless stream of promissory value. Cooper singles out this nonconsonant logic in observing that "the crude truth of capitalism, according to Marx . . . is that in the long run, there is no mediation to be found between the rhythms of production and consumption, no progressive transition from third to second to first world that does not at some point enter into conflict with the blunt need to increase the production of relative surplus value."[56] If one meaning of *resolution* refers to the harmonization of formally incompatible notes, an irresolvable tension emerges in this incarnation of global health: the structural impossibility of so-called second and third worlds becoming synchronized, through the raising of capacity, with global standards of health set out by the first. Commonality, sharing, mutuality, oneness, must simultaneously be mobilized and undermined in order for emerging microbes to function as the promising basis of the new and horizonless economic frontier.

Indeed, it was this very mechanism that prolonged the Covid-19 pandemic in 2020. After the largest multinational pharmaceutical corporations such as Pfizer and Moderna had developed and produced effective vaccines against Covid-19, they eventually sold large volumes to most nations in the developing world.[57] In the interim, many developing countries were being crushed under wave upon wave of infection because they were unable to access the new vaccines—the most crucial element in bringing the pandemic under control. While talk about vaccine "shortages" abounded, this artificial scarcity was rooted in an informational infrastructure that the TRIPS agreement was designed to sustain. More than one hundred vaccine manufacturers around the world were ready and waiting to purchase licenses,[58] yet they needed an intellectual property (IP) release before proceeding to ensure they would not be punished by the WTO for producing generic equivalents. In October 2020, led by South Africa and India and backed by the Africa Group, Least-Developed Countries Group, and nineteen other member states, a TRIPS waiver was requested for the Covid vaccines.[59] This waiver would have enabled other manufacturers to produce vaccines and meet supply needs worldwide. When President Biden finally agreed to back the waiver in May 2021 (crucially, supporting the release on vaccines but not on related technologies, techniques, and treatments),[60] Pharmaceutical Research and Manufacturers of America (PhRMA) decried the decision in the name of defending "innovation." The TRIPS waiver, they claimed, would "foster the proliferation of counterfeit drugs" and erode the U.S. economy by "handing over American innovations to countries looking to undermine our leadership in biomedical discoveries." Moreover, it would "cede American innovation leadership to countries like China."[61] Meanwhile, the EU blocked the release by offering to emplace compulsory licensing agreements instead of an actual waiver. It claimed that compulsory licensing would help to expand vaccine production by enabling countries to purchase licenses for some components from various rights holders.[62] Pfizer's and Moderna's vaccines require around 280 ingredients sourced from at least nineteen countries, and these companies have formulation, production, and packaging factories in still more countries.[63] The EU agreement required producers to obtain licenses for proprietary information all along the manufacture and supply chain, which would take years. Moreover, compulsory licensing did not cover other proprietary information required to make use of patented components of the vaccine, from clinical trial data and packaging designs to copyrighted manuals and protein recipes.[64] Unsurprisingly, the waiver block was supported by an array of multinational pharmaceutical corporations, government bodies, and lobbyists. The WTO itself was to work toward a decision in December 2021, but was (ironically) compelled to postpone the meeting for a second time due to the emergence of the Delta variant, which surfaced in part due to the unchecked viral spread of the original strain. While most developed countries boasted high vaccination rates (on average above 80 percent) and the United States possessed an excess of vaccines at the time, developing countries were still struggling to reach a 10 percent vaccination rate.

By March 2023, discussions on the TRIPS waiver languished in the bureaucracy of the WTO. Meanwhile, the pandemic has caused close to six million deaths worldwide as variants of the virus continue to emerge. While this fight over the price of knowledge and the flow of information effectively sets a price on life, the same large pharmaceutical corporations that blocked the waiver are developing new drugs, formulations, and treatments to respond to the evolving virus—its mutation to the omicron variant and subvariants.[65] In this context, IP attorney Tahir Amir observes that "by kicking the can down the road, the Pfizers and Modernas can get more efficient and more to scale, and they'll say, 'Oh look, the current supply is going to meet demand.'"[66] This deliberate orchestration of informational drag and flow sets rhythms of production and consumption that "ensure the right amount of product in pipelines," to cite *America's Vital Interest*'s prescient and prescriptive formulation once more. A full accounting of the Covid-19 pandemic is not yet possible, but future analyses will need to account for the longer history of international patent protections and the global health market.

The Eternal Bonds of Emergent Risk

Tom Cohen observes that "the Ponzi-scheme logics" of the twenty-first-century render an earthscape of "catastrophic deferrals" in "a voracious present that performs a kind of tempophagy [time-eating] on itself."[67] The project of raising capacity is governed by this logic of catastrophic deferral. It generates economic value out of the vagaries of emergent risks, and its framework of expansion hedges capital production against the very futures that biosecurity purports to secure. Its economic productivity counts on crisis. Nowhere is this catastrophic logic more visible than in Wolfe's own trajectory. Once a prominent AIDS researcher, he then transformed that laboratory into the headquarters for GVFI in partnership with its newly opened sister company, Metabiota. The latter is a for-profit firm that sells pandemic risk analysis to enable insurance and reinsurance corporations to underwrite and finance risk. In 2020, GVFI closed, while Metabiota prospered. In keeping with the most extreme version of preparedness discourse, the company that survived focuses not on treatment or building infrastructure to address present dangers; Wolfe's goal is not to make vaccines or other pharmaceuticals to combat emerging microbial threats, although he advocates for their worthiness. What, then, is the finished product Metabiota is working towards? To build analytic capacity, to grow the power of remunerative prediction.[68]

Wolfe argues that emerging futures can be rendered more resilient by deploying predictive technologies to help the world prepare for coming crises. Crucially, Metabiota's success does not rely on its ability to predict with accuracy which pandemic threats will arrive on our doorstep. The informational resources of GVFI became the basis for the highly speculative and highly profitable risk-based analysis that Metabiota deploys to bring emergent and potentially catastrophic risks into sight as calculable, securable, and commodifiable assets.[69] Echoing the dynamic by which the innovation economy encloses the inexorabilities of change to make value ex nihilo, Wolfe describes his work at Metabiota as follows: "There is a bit of financial alchemy to the whole thing. You are really creating something from nothing."[70] Risk and speculative finance are key markers of this approach to emerging microbial threats;[71] they blur and span the lines between assets and liabilities while fashioning risk itself as productive of value.[72] The regulatory environment incentivizes this structure by transmuting catastrophic threat into an asset to be bought and sold. As a result, nature's creative potentiality is effectively harnessed to a system that endlessly writes and rewrites risk back across the planet. This self-propagating logic is visually echoed in Metabiota's corporate

Figure 5.1. Screen capture of Metabiota's company logo, metabiota.com, 2019.

logo (Figure 5.1). GVFI's corporate logo revealed its vision of the globe bound together by geometric lines, a networked space pervaded by common risk; in Metabiota's logo, those lines of planetary threat morph into a globe made up of concentric rings of arrows pointing back into themselves in an endless circuit of productivity. At their heart, these rings both form and enclose a black hole: an inadvertent, dark reminder of the places set in continual crisis from which cycles of productivity are extracted.[73] This logo articulates the structural impossibility of a global health project that, in line with Cohen's tempophagy, feeds back on itself and shunts crisis to the future. The world bound together by shared risk is a globe linked in obligation to a harmonized response, and now tied to the eternal circuity of a self-perpetuating logic.

What feeds this eternal circuit? If the bioeconomy is a product of the United States' search for new and unlimited resources, the limitless threat of microbes provides the answer: emerging microbes are pure production. Here, the concept of microbial emergence reveals one of its most potent effects: the self-regenerating structure of novel viruses—their capacity to emerge at any time, in any place, and in inexhaustible genetic combinations—promises boundless economic production that speculates on the possibility of perpetually infected futures.[74] By this logic, the point is not to halt microbial emergence, nor even to mitigate it. The goal is simply to manage the constantly unfolding threshold of emergence by directing its pluripotency to unlimited and endlessly productive resolutions.

Wolfe articulates this bioeconomic logic in a spirited speech to prospective microbiology students at Stanford University. He asserts that we would be mistaken in believing that the "great age of exploration on earth is over," because there are unknown microbial worlds of "dark matter" that teem with "incredible potential." In a later version of this talk, Wolfe explains why microbes constitute the next great frontier, aligning GVFI's work with a broader story of Western colonial exploration and expedition by evoking the celebrated conquests of the past. His framing of microbiology draws on this colonialist narrative of frontierism in which new territories are discovered and claimed. "We've mapped all the continents," he says; "What's left to explore?" His answer is microbes. Wolfe's "dark matter" not only references the unseen world of physics; it also strongly echoes the all-too-familiar colonial conquests in which "dark continents" are "discovered," mapped, and claimed. Do not mourn the end of exploration, he offers, for frontiers are unfolding everywhere; "There are unknowns all around us, and they're just waiting to be discovered."[75]

Our current model of global health is premised on the continued colonial quest to explore the endless frontiers of emerging microbes. If the systemic injustices of both postcolonial and settler-colonial orders underwrite the health and might of wealthy Western nations, then a global health project that operates on preemptive biopreparedness might better be termed "ghoul health," in Elizabeth Povinelli's phrase: a condition haunted by the "bad faith of liberal capital and its multiple geophysical tactics and partners."[76] The logistical workings of the bioeconomy ensure that ghoul health is the logical endpoint of its reliance on a system that produces endless value. As the dark heart of a militarized and neoliberal global health, there are consequences of perpetuating crisis in those "elsewheres." Microbiologically, this "offshoring" of crisis inevitably comes "home" to roost not only in the global flow of newly mutant microbes. The war against microbial emergence also haunts as its effects become physically registered in planetary life. Its project suffuses the world with the mutant temporalities of the war against emerging microbes and its deferred consequences, sped-up futures, and its borrowing of time from certain places of the world for specific beneficiaries. Ghoul health in fact creates the planetary conditions for new forms of biovirulence, and all futures are haunted by it. If microbial resolution posits this ever-possible, always looming threat of biospheric karmic materialization, might this effort to anticipate potential viral futures reflect a growing awareness that unpayable anthropogenic debts are quickly coming due?

Acknowledgments

It is impossible to trace out the many key moments that shape and reshape a book. There is no single moment that marks when the idea began, or a clear trajectory of how and why it transformed through the years. This work developed out of its enmeshment in the most felicitous clouds of conversations, seminars, institutions, friendships, and interlocutors that made the conditions of its thinking and writing possible. While it is difficult to cleanly delineate these debts, I will document them as best I can. So many have offered encouragement, feedback, and advice over the years, yet, all of the shortcomings and inaccuracies in this book are mine alone.

The research and writing of this book was made possible with generous funding and support from the American Council of Learned Societies, the Susan B. Anthony Research Institute, the Andrew Mellon Foundation, and University of California Humanities Research Institute.

The early seeds of this book were planted when I was a PhD student in the Graduate Program for Visual Culture at the University of Rochester. It was a saving grace to be among peers and faculty there who reminded me that my brain did not exist in a state of incoherent chaos; instead, the term for this seemingly cluttered assembly of thoughts is "interdisciplinarity." A group of people taught me how to write and think in this mottled terrain: Sharon Willis knew when to press and when to give one space. I miss conversations shared with Robert Foster over drinks at Solera and often left with glimmering notions, many of which evolved into some of the thoughts in these pages. A. Joan Saab was always ready with sage advice and a keen eye. Theodore Brown, with the wisdom of a historian, taught me how to temper my claims about the contemporary. Maureen Gaelens, Marty Collier, Cathy Humphrey, and Donna Derks dotted all the administrative i's and crossed the bureaucratic t's to ensure that all went smoothly. The research librarians and staff at the university's library remain unmatched, with special thanks to all working at the Art and Music Library, and particularly to Head Librarian Stephanie Frontz: a stellar librarian, but also a caring landlord and a dear friend. This project also benefited from conversations with the wonderful colleagues I met during my postdoctoral fellowship at Hobart and William Smith and the Central New York Humanities Corridor meetings for the medical humanities.

When I set off to the Midwest for the Provost's Postdoctoral Fellowship at the Center for 21st Century Studies at the University of Wisconsin–Milwaukee, I could not have anticipated just how critical that would be for this project, nor what the community there would come to mean to me. I owe much to the fellow fellows with whom I was fortunate to spend a year reflecting on the theme of "Humanities Futures." I deeply appreciate Richard Grusin's directorship at the Center during my time there, his sharp intellect, and the conversations shared over countless walks and through times of hilarity and heartbreak. Ivan Ascher, Aneesh Aneesh, Emily Clark, Elena Gorfinkel, Jennifer Johung, Elana Levine, Jenna Loyd, Jesse McAfee, Annie McLanahan, Tasha Oren, Maria Peeples, Anne Pycha, Jason Puskar, Alison Sperling, Mark Vareschi, Tami Williams—such loving thanks to them all for housing, conversations, camaraderie, and so much more. Much appreciation to those who gathered at my place for monthly "No Miracles. Just Soup," where we warded off the pain of Midwest winters and supported one another through a politically tumultuous era for the University of Wisconsin system. My Milwaukee people showed me that those oft-mocked "bland flatlands" are neither bland nor even topographically flat, but full of energy and dimension, and crackling with intellectual and political life.

This book found its next home at Cornell University when I was a fellow at the Society for the Humanities/Cornell Atkinson Center for Study of Sustainability. I remain grateful to Tim Murray for his astute directorship at the Society, for being so brilliant and gracious, and for having a ridiculously enormous heart. The cohort of fellows there provided an unmatched environment for the most fun and exciting intellectual gymnastics. I was always eager to get to our weekly seminars where we pushed one another to the edges of thought (and often straight into the untamed realm beyond). Heartfelt thanks to all of them for sharing their work and their feedback and for being so exciting to think with. I feel the presence of the conversations we shared throughout this book. Key aspects of this book were also formed from regular after-hours conversations and shenanigans with particular fellows: Gemma Angel, Naminate Diabate, Pamela Gilbert, Alicia Imperiale, Stacey Langwick, Karmen MacKendrick, Emily Rials, Elyse Semerdjian, Samantha Sheppard, Nancy Worman, and Seçil Yilmaz. I still can't believe how lucky we were (and are) that the universe dared to stuff so many strange, wonderfully mischievous, and blisteringly smart oddballs into the A.D. White House at the same time. A faculty residential fellowship at the Cornell Branch Telluride House made it possible for me to live and write in Ithaca with ease. It was a privilege to spend time in a community of such intellectually engaged folks. I can't say how much I appreciated decompressing and commiserating about academia with Rumit Singh and Kris Hartley, the two other faculty fellows; most crucially, though, I'm sure it was our side-splitting nighttime antics that really kept us all out of the deep end.

A group of kin has been around this book (and around me with this book) for so long, and they have contaminated it (and me) in the most magical ways. Aubrey Anable has been holding space (and other things) for me ever since she rescued me from being lost in the basement of Rush Rhees Library. Vicky Pass has stood by me, cheek to cheek, ever since we first tangoed in an abandoned subway tunnel in Rochester, New York. That city is also where I first met Eleana Kim, and now we are neighbors in Long Beach, California; between the Atlantic and the Pacific Oceans, the northeast snow and So Cal sun. . . . There's an unwieldy metaphor somewhere in there that expresses how elemental her friendship and intellect have been for this book, so maybe it's simplest to say "Thanks for everything." Frank Straka nurtured this book, and me, with many years of love, conversations, and beautiful walks. He unwaveringly cheered me on and stood by me through the many ups and downs of academic life. And he made sure that good music was playing and the floors were cleared when there was nothing left to do but to dance it all out. Imani Kai Johnson regularly reminds me to focus on the weirdest parts of my writing (and my life) when I've lost my voice, because that's where I'll find myself again (although I'm pretty sure I found her there, too). I sometimes wonder if Brian Grosskurth's absent presence might have inflected this book in some way from the start: through him, in another life, I first began thinking about the riddles of appearance. More recently, he read every word of this book with love, named words that hovered stubbornly on the tip of my tongue, and demonstrated a saint-like patience with me as I put him through my twisted (and argumentative) process of rethinking, reformulating, and rewriting.

For their friendship, advice, time, encouragement, conversations, and emotional and literal nourishment through the years I thank Crystal Myun-Hee Baik, Katherine Behar, Janet Berlo, Lisa Bhungalia, Elaine Brindley, Becky Burditt, Jenny Chung, Tom Clifford, Lyell Davies, Heather Davis, Aviva Dove-Viebhan, Michael B. Gillespie, Amanda Graham, Alicia Iñez Guzman, Jessica Hayes-Conroy, Dinah Holtzman, Emily Hue, the Lambert and Bowerman clan (Deborah, Lisa, and Tashi), Howard Lonn, Nicola Mann, Xhercis Mendez, Julian Nykolak, Claire Sagan, Nic Sammond, Leah Shafer, SA Smythe, Tess Takahashi, Lisa Uddin, Jane Ward, and Iskander Zulkairnain.

Much of this book was developed during the eleven (and counting) annual summer writing retreats with the O.G.s and other friends who have joined us through the years. Not only incredibly productive, these retreats have shown us how to cultivate a more graceful and pleasure-filled relationship with our work, what we do, and what we do to ourselves (and to one other) in it.

I thank my brilliant colleagues in the Department of Media and Cultural Studies at the University of California, Riverside; I'm truly lucky to be part of such a vibrant department. A special thanks to Judith Rodenbeck for being an incredible chair, and for supporting me and this book in countless ways. I am fortunate to have Sherryl Vint as intellectual kin and mentor. Diana Marroquin, Irene Dotson, and all the financial and administrative staff make everything possible; my gratitude to them for lighting a path through a maze of red tape, technological glitches, and paperwork, so I could move my research and writing forward. This book benefitted enormously from a 2020–21 University of California Humanities Research Institute Presidents' Faculty Research Fellowship, which enabled me to focus on writing for a full year.

Three readers for a manuscript workshop—Ivan, Eleana, and Aubrey—cared enough not to mince their words and offered their feedback to improve the first draft of this book. I am immensely grateful to Annie Moore (themagicword.ca) for her editing; no doubt I (and readers of this book) would be stuck in dizzying whirlpools of verbiage, or left wandering in mazes of thought, were it not for her sensitive and smart work. Kristina Wilson lent me her expertise on gene sequencing techniques and optical physics in conversations always beautifully inflected with her romantic soul. Lisa Sutcliffe, then curator at the Milwaukee Art Museum, invited me to say a few things about Trevor Paglen's work at MAM, which gave me an opportunity to reflect on the relevance of his work to my argument in chapter 3. Thanks to Mark Vareschi and the students in his graduate course on data cultures at UW–Madison, who invited me to the Center for the Humanities to share my initial ideas on birds as data visualization technologies. Nicole Starosielski offered her comments on parts of this manuscript as well as her enthusiastic support. I'm grateful to her for inviting me to reflect on microbes as elemental media with her graduate class at New York University and to the students in that course for their smart and generative questions.

I thank Doug Armato for his sound guidance as this book moved through various stages of writing and development. Zenyse Miller was on top of all the important details as this book moved toward production. Two reviewers, Stefanie Fishel and the other anonymous, helped sharpen the manuscript's contribution, and I thank them for their discerning advice and comments. I feel so fortunate that David Cecchetto and Arielle Saiber invited me to publish this book with the Proximities series; they have my gratitude for making space for scholarly experiments in nearness.

The cover image is from a series of paintings made by Sarah Roberts, Dr. Simon Park, and the bacteria *Serratia marcescens*. I thank Roberts and Park for making these works available through

Creative Commons. XiaoXiao Li graciously took in my ideas and requests as we worked on modifications to the cover image and I so enjoyed learning about the technical aspects of resolution with her. Wade Keye expertly helped me with image formatting. David Martinez was a master indexer. Rachel Moeller has been an incredible and patient production manager.

Parts of this book were written during the Covid-19 pandemic. I have distinct memories of trying to research and write while navigating a cross-country relocation (and eventually, one to another country) amid the intensity of its first several months. A move like that is difficult in the best of times; attempting one in a time of panic, curfews, lockdowns, and border closures requires a finely tuned logistical ballet. I am painfully aware that I would have lost many months of writing were it not for a team of friends, relatives, and the occasional stranger, who all pitched in to help so I could keep working through maximum instability. Frank did a lot of the emotional and physical heavy lifting; Jesse gave up his studio for three months when I didn't have any place to work; Patrick Keilty loaned me office furniture; Nic saved me when I could not access many crucial books; Lisa Bhungalia and Alexis Salas helped me think through some planning twists and turns; a stranger gave me her spare book bag when she noticed that mine broke in transit; Esther Lee and Mi-Jung and David Gast did so much more than cousins should ever have to do. Among many other things, they helped route my belongings from one end of the country to the other and prepared my getaway car.

My biggest debt is to my hilarious and loving family. I owe them everything absolutely. My mind explodes with overload when I try to recount the uncountable ways that Grace and Peter Yoo so generously supported me. Naomi, Jessica, and Brody took me on walks and talks and inspired me with their weirdness and beautifully curious spirits. Iun Sook Kim and Sung Rack Kim gave me more than I can ever name, let alone repay. They also trusted and supported their youngest daughter even though they couldn't understand what this thing called academia was, or why one would choose to be in it. And they put on their bravest faces when they had to explain to their friends that, while very much an adult, she was "still in school."

This book is dedicated to a creature who will never read its words, but who may one day decide on a whim to taste its pages (especially the one that mentions the word "chicken"): Sadie Pickles, my beloved microbiomial kin, kindred spirit, and partner in all things. Whatever lucky stars aligned to bring us together, let's smell them all. Everything is for you.

Notes

Introduction

1. This 1989 meeting was sponsored by the U.S. National Institutes of Health. It was convened by Stephen Morse, then faculty at Rockefeller University and advisor at the World Health Organization (WHO); Joshua Lederberg, then president of Rockefeller University, Nobel laureate, and WHO advisor; and Donald A. Henderson, former Officer of the U.S. Communicable Disease Control (now Centers for Disease Control and Prevention). The conference was attended by more than two hundred participants.

2. René Dubos, *Mirage of Health: Utopias, Progress, and Biological Change* (New Brunswick, N.J.: Rutgers University Press, 1959).

3. For more on the ways that the development of microbial "counter-proliferation technologies" helped to frame human dealings with microbes in the terms of war, see IOM, *Ending the War Metaphor: Changing Agenda for Unravelling the Host-Microbe Relationship* (Washington, D.C.: National Academies Press, 2006).

4. Joshua Lederberg, "Viruses and Mankind: Intracellular Symbiosis and Evolutionary Competition," *Frontline,* Public Broadcasting System, accessed March 4, 2023, https://www.pbs.org/wgbh/pages/frontline/aids/virus/humankind.html.

5. Paul Farmer, *Infections and Inequalities: The Modern Plagues* (Berkeley: University of California Press, 1999), 39.

6. On the accompanying shift in public health regimes, see Andrew Lakoff, "Two Regimes of Global Health," *Humanity: An International Journal of Human Rights, Humanitarianism, and Development* 1, no. 1 (2010): 59–79, https://doi.org/10.1353/hum.2010.0001; Nicholas King, "Security, Disease, Commerce: Postcolonial Ideologies of Global Health," *Social Studies of Science* 32, no. 5/6 (2002): 763–89. On the connections between preemptive biopreparedness and the security paradigms of the Cold War, see Elizabeth Fee and Theodore Brown, "Preemptive Biopreparedness: Can We Learn Anything from History?" *American Journal of Public Health* 91, no. 5 (2001): 721–26.

7. The development of the emerging microbe concept was accompanied by another conceptual shift in U.S. public health discourse, one marked by a move in terminology, from "international health" to "global health." The term *global health* was not commonly used prior to this time. Whereas the former reflects the idea that disease threats should be managed between individual states, the latter emphasizes the new idea that such matters are global problems requiring the development of global governance projects geared at addressing emerging infections as matters of biospheric significance. See Theodore Brown, Marcus Cueto, and Elizabeth Fee, "The World Health Organization and the Transition from 'International' to 'Global' Health." *American Journal of Public Health* 96, no. 1 (2006): 62–72.

8. IOM Committee on Emerging Microbial Threats to Health, *Emerging Infections: Microbial Threats to Health in the United States,* ed. Joshua Lederberg, Robert E. Shope, and Stanley C. Oaks Jr. (Washington, D.C.: National Academies Press, 1992).

9. On the revitalization of Cold War institutional relations under President Bill Clinton and after, see David P. Fidler, "Public Health and National Security in the Global Age: Infectious Diseases, Bioterrorism, and Realpolitik," *George Washington International Law Review* 35, no. 4 (2003): 787–856; Judith Miller, Stephan Engelberg, and William Broad, *Germs: Biological Weapons and America's Secret War* (New York: Simon & Schuster, 2001). For an analysis of the connections between the contemporary regimes of public health security and Cold War biopreparedness frameworks, see Fee and Brown, "Preemptive Biopreparedness: Can We Learn Anything from History?"

10. See Brown, Cueto, and Fee, "The World Health Organization and the Transition from 'International' to 'Global' Health." On the expansion of the emerging infections concept to several other countries and Canada, and the UN revisions to the International Health Regulations, see Lorna Weir and Eric Mykhailovskiy, *Global Public Health Vigilance: Creating a World on Alert* (New York: Routledge, 2010).

11. Theories of nonlinear dynamics were studied in the late nineteenth century. For an overview of their development into chaos theory, and ensuing branches and movements, see N. Katherine Hayles, "Introduction: Complex Dynamics in Literature and Science," in *Chaos and Order: Complex Dynamics in Literature and Science,* ed. N. Katherine Hayles (Chicago: Chicago University Press: 1991).

12. The term "the science of chaos" is often used interchangeably with "complex systems science," although such systems are not believed to be completely random and absent of order. I use both terms here to maintain a view of the scientific and cultural stakes of investigating systems that defeat usual methods of analysis.

13. One might also note that in one strand of twentieth-century philosophy, questions of emergence as a coming-to-presence is a central preoccupation. See, for example, Maurice Merleau-Ponty, *The Visible and the Invisible* (Evanston, Ill.: Northwestern University Press, 1968); Jean-Luc Nancy, *The Birth to Presence* (Stanford, Calif.: Stanford University Press, 1994); Giorgio Agamben, *Potentialities* (Stanford, Calif.: Stanford University Press, 1999).

14. Rather than apprehending the world as entirely absent of order, this science restructures the relation between chaos and order, holding that within the disorder of complex systems lay a hidden, deep structure. I return to this briefly in chapter 1.

15. For an account that historically situates such militaristic frameworks for microbial risk, see Fee and Brown, "Preemptive Biopreparedness." The war metaphor was cemented in Lederberg's 1989 speech (see note 3). See also Joshua Lederberg, "Microbiology's World Wide Web," *Project Syndicate,* December 1, 2000, https://www.project-syndicate.org/commentary/microbiology-s-world-wide-web. It is also important to note that the IOM hosted a workshop on the problems of using war metaphors in matters of microbial governance. The workshop committee argued for more symbiotic metaphors to cultivate a more dynamic, coevolutionary understanding of microbes. The war metaphor, however, proves durable: while it never died away, and as pandemics like Zika (2016), Ebola (2015), swine flu (2009), and Covid-19 (2020) have shown, it was also the most available and common framework that experts and laypersons fall back on during times of heightened microbial risk. In my analysis, moreover, it is necessary to ask how such calls to end the war metaphor take part in the broader ecologization of microbial warfare that I trace in this book.

16. IOM, *Emerging Infections,* 137.

17. David Satcher, "Emerging Infections: Getting Ahead of the Curve," *Emerging Infectious Diseases* 1, no. 1 (1995): 1–6.

18. Priscilla Wald, *Contagious: Cultures, Carriers and the Outbreak Narrative* (Durham, N.C.: Duke University Press, 2008); Kristen Ostherr, *Cinematic Prophylaxis: Globalization and Contagion in the Discourse of World Health* (Durham, N.C.: Duke University Press, 2005).

19. For an illuminating reading of Steyerl's video, and an analysis of resolution in relationship to the scales of information in the digital age, see Aubrey Anable, "The Work of Didactic Art in the Information Age," *ASAP Journal* 6, no. 1 (2021): 51–58.

20. Eyal Weitzman, *Forensic Architecture: Violence at the Threshold of Detectability* (Brooklyn, N.Y.: Zone Books, 2017).

21. For example, Weitzman notes that U.S. public satellite image providers adhere to the 0.5 meter rule, except in Israel and Palestinian territories. Weitzman, *Forensic Architecture,* 29.

22. Weitzman, *Forensic Architecture.*

23. For histories of resolution, see Laura Kurgan, *Up Close and at a Distance: Mapping, Technology, Politics* (Brooklyn, N.Y.: Zone Books, 2013); Alvy Ray Smith, *A Biography of the Pixel* (Cambridge, Mass.: MIT Press, 2021). For a remarkable analysis of grids as possible precursor to resolution grids, see Phillip Thurtle, *Biology in the Grid: Graphic Design and the Envisioning of Life* (Minneapolis: University of Minnesota Press, 2018).

24. Ludwig Fleck, *Genesis and Development of a Scientific Fact* (Chicago: University of Chicago Press, 1979).

25. Barbara Herrnstein-Smith, *Scandalous Knowledge: Science, Truth and the Human* (Durham, N.C.: Duke University Press, 2006), 56.

26. Timothy Lenoir, *Instituting Science: The Cultural Production of Scientific Disciplines* (Stanford, Calif.: Stanford University Press, 1997), xxi.

27. The United States' assessed annual contributions to the WHO in the mid-1980s made up 25 percent of the agency's annual operating budget (estimated at US$73 million). By slowing or stopping membership payments, or by withdrawing its membership, the United States could create severe budgetary constraints for the agency. The United States has used its economic might to put pressure on the WHO to influence its programs and agendas. These strategies enabled the United States to, for example, advance the interests of its pharmaceutical companies in 1985; in 1989, the United States stated that they would withdraw from the WHO if the agency admitted the Palestinian Liberation Organization and an independence movement in Namibia (SWAPO) as member states. See Marcus Cueto, Theodore Brown, and Elizabeth Fee, *The World Health Organization: A History* (Cambridge: Cambridge University Press, 2020), 243–46.

28. See Cueto, Brown, and Fee, *The World Health Organization,* 203–61.

29. Cueto, Brown, and Fee, *The World Health Organization,* 285.

30. While the logic of "preemptive biopreparedness" seems, at first glance, to have surfaced as part of President Bush's preemptive war, its logic draws from paradigms of Cold War biodefense. See Fee and Brown, "Preemptive Biopreparedness."

31. Precautionary principles share certain traits with preemption and preparedness; all are modes of governing potential futures brought about through global interconnectedness. See Ben Anderson, "Precaution, Preemption, Preparedness: Anticipatory Action and Future Geographies," *Progress in Human Geographies* 34, no. 6 (2010): 777–98. For a reading of the commonalities of the precautionary and preemptive logics, see Marieke de Goede and Samuel Randalls, "Precaution, Preemption: Arts and Technologies of the Actionable Future," *Environment and Planning D, Society and Space* 27, no. 5 (2009): 859–78.

32. Melinda Cooper, *Life as Surplus: Biotechnology and Capitalism in the Neoliberal Era* (Seattle: University of Washington Press, 2008), 82.

33. See Andrew Lakoff, *Unprepared: Global Health in a Time of Emergency* (Oakland: University of California Press, 2017); Lindsey Thomas, *Training for Catastrophe: Fictions of National Security* (Minneapolis: University of Minnesota Press, 2021); Anderson, "Precaution, Preemption, Preparedness."

34. See Andrew Lakoff, "The Generic Biothreat, or, How We Became Unprepared," *Cultural Anthropology* 23, no. 3 (2008): 399–428.

35. On the broader history of the dovetailing of health and security see, for example, Andrew Lakoff, *Unprepared*; Victor W. Sidel, Hillel W. Cohen, and Robert M. Gould, "Good Intentions and the Road to Bioterrorism Preparedness," *American Journal of Public Health* 91, no. 3 (2001): 716–18; Nicholas King, "The Influence of Anxiety: September 11, Bioterrorism and the American Public Health," *Journal of the History of Medicine* 58 (2003): 433–41; Nancy Tomes, "The Making of Germ Panic: Then and Now," *American Journal of Public Health* (2000): 90, 191–98.

36. Mike Leavitt, "Pandemic Preparedness: Thinking the Unthinkable," speech at the National Press Club, Washington, D.C., October 27, 2005, https://www.c-span.org/video/?189600-1/pandemic-preparedness-thinking-unthinkable (also cited in Lakoff, "Generic Biothreat").

37. Joseph Masco uses this term in relation to the ruined futures posited by the nuclear bomb. Joseph Masco, *The Nuclear Borderlands: The Manhattan Project in Post–Cold War New Mexico* (Princeton, N.J.: Princeton University Press, 2006).

38. For analyses of risk in relation to such topics, see, for example, Jason Puskar, *Accident Society: Fiction Collectivity and the Production of Chance* (Stanford, Calif.: Stanford University Press, 2012); James Tulloch and Deborah Lupton, *Risk and Everyday Life* (London: Sage, 2003); Louise Amoore, "Biometric Borders: Governing Mobilities in the War on Terror," *Political Geography* 25, no. 3 (2006): 336–51.

39. Ulrich Beck, *Risk Society: Towards a New Modernity* (London: Sage, 1992).

40. See François Ewald, "Insurance and Risk," in *The Foucault Effect: Studies in Governmentality,* ed. Graham Burchill, Colin Gordon, and Peter Miller (Chicago: University of Chicago Press, 1991); François Ewald, *The Birth of Solidarity: The History of the French Welfare State,* trans. Timothy Scott Johnson (Durham, N.C.: Duke University Press, 2020); Frank Knight, *Risk, Uncertainty, and Profit* (Boston: Houghton Mifflin, 1921).

41. Such a field involves the pervasive awareness that one might be disastrously and intimately impacted in "unforeseeable," "incalculable,"

and "immeasurable" ways by anyone, anywhere, within the world. Beck, *Risk Society,* 180, 170, 178.

42. N. Katherine Hayles, *How We Became Posthuman: Virtual Bodies in Cybernetics, Literature and Informatics* (Chicago: University of Chicago Press, 1999); Orit Halpern, *Beautiful Data: A History of Vision and Reason since 1945* (Durham, N.C.: Duke University Press, 2014); Lily Kay, *Who Wrote the Book of Life: A History of the Genetic Code* (Stanford, Calif.: Stanford University Press, 2000).

43. Hayles, *How We Became Posthuman*; Arthur Kroker and Marilouise Kroker, *Hacking the Future: Stories for the Flesh-Eating 90's* (New York: St. Martin's, 1996).

44. Marshall McLuhan, *Understanding Media: The Extensions of Man* (Cambridge, Mass.: MIT Press, 1964); Rahul Mukherjee, *Radiant Infrastructure: Media, Environment and Cultures of Uncertainty* (Durham, N.C.: Duke University Press, 2020).

45. Bishnupriya Ghosh, "Becoming Undetectable in the Chthulucene," in *Saturation: An Elemental Politics,* ed. Melody Jue and Rafico Ruiz (Durham, N.C.: Duke University Press, 2021). See also Bishnupriya Ghosh, *The Virus Touch: Theorizing Epidemic Media* (Durham, N.C.: Duke University Press, 2023).

46. Eugene Thacker, *Biomedia* (Minneapolis: University of Minnesota Press, 2004).

47. Jason Moore, *Capitalism and the Web of Life: Accumulation and the Web of Capital* (New York: Verso, 2015); Rob Nixon, *Slow Violence and the Environmentalism of the Poor* (Cambridge, Mass.: Harvard University Press, 2011); William Connolly, *Facing the Planetary: Entangled Humanism and the Politics of Swarming* (Durham, N.C.: Duke University Press, 2017).

48. In this vein, Anne Pasek, Yuriko Furuhata, and Heather Davis analyze the microparticulate and molecular dimensions of such transformations. See Anne Pasek, "Carbon Vitalism: Life and the Body in Climate Denialism," *Environmental Humanities* 13, no. 1 (2021): 1–20, https://doi.org/10.1215/22011919-8867175; Heather Davis, *Plastic Matter* (Durham, N.C.: Duke University Press, 2022); Yuriko Furuhata, *Climactic Media: Transpacific Experiments in Atmospheric Control* (Durham, N.C.: Duke University Press, 2022).

49. Richard Dyer, *White: Essays on Race and Culture* (Abingdon, UK: Routledge, 1997).

50. See, for example, Zakiyyah Iman Jackson, *Becoming Human: Matter and Meaning in an Antiblack World* (New York: New York University Press, 2020); Mel Y. Chen, *Animacies: Biopolitics, Racial Mattering and Queer Affect* (Durham, N.C.: Duke University Press, 2012); Amy Bhang, *Migrant Futures: Decolonizing Speculation in Financial Times* (Durham, N.C.: Duke University Press, 2017).

1. Pathogenic Nation Making

1. *Deadly Migration* was also available via clickable banners in online publications of newspapers, such as the *New York Times,* and on IBM's YouTube channel.

2. Joseph Masco, *Theater of Operations: National Security Affect from the Cold War to the War on Terror* (Durham, N.C.: Duke University Press, 2014), 7.

3. Although terror, anxiety, and fear are not American emotions per se, Masco argues that they are renationalized and mobilized in the American counterterror state. As such, he argues that the counterterror state (and its various articulations in areas such as public information, weather forecasting, and biosecurity) sits in continuity with "a specific American logic and domestic history" stemming from the Cold War. Masco, *Theater of Operations,* 3.

4. Masco, *Theater of Operations,* 37.

5. Priscilla Wald, *Contagious: Cultures, Carriers and the Outbreak Narrative* (Durham, N.C.: Duke University Press, 2008), 6–14.

6. Wald, *Contagious,* 29–53.

7. See Melissa Gregg and Geoffrey J. Seigworth, "An Inventory of Shimmers," in *The Affect Studies Reader,* ed. Melissa Gregg and Geoffrey J. Seigworth (Durham, N.C.: Duke University Press, 2009), 1–27.

8. I will return to discuss the relationship between the wild, the domesticated, and the racialization of the future biological catastrophe in chapter 3.

9. Lindsey Thomas, *Training for Catastrophe: Fictions of National Security* (Minneapolis: University of Minnesota Press, 2021), 15, 81.

10. Anthony Giddens, *The Consequences of Modernity* (Stanford, Calif.: Stanford University Press, 1990), 140.

11. Robert Foster, *Coca-Globalization: Following Soft Drinks from New York to New Guinea* (New York: Palgrave Macmillan, 2008), 19.

12. Jennifer Brower and Peter Chalk, *The Global Threat of New and Re-Emerging Diseases: Reconciling US National Security and Public Health Policy* (Santa Monica, Calif.: RAND Science and Technology, 2003), viii.

13. David Satcher, "Emerging Viruses; Getting Ahead of the Curve," *Emerging Infections* 1 no. 1 (1995): 1–6.

14. Institute of Medicine, *Orphans and Incentives: Developing Technology to Address Emerging Infections* (Washington, D.C.: National Academies, 1997).

15. Speech delivered by Pres. Clinton during the 2003 Kennedy Presidency Lecture.

16. Louise Amoore, *The Politics of Possibility: Risk and Security Beyond Probability* (Durham, N.C.: Duke University Press, 2013).

17. From Mike Leavitt, "Pandemic Preparedness: Thinking the

Unthinkable," speech at the National Press Club, Washington, D.C., October 27, 2005, https://www.c-span.org/video/?189600-1/pandemic-preparedness-thinking-unthinkable.

18. The laying out of "worst-possible scenarios" and thinking possible futures through catastrophic and anticipatory frameworks has become a common strategy by which experts seek to make knowable the futures of EIDs. For sustained critical discussions of this technique, see Andrew Lakoff, "The Generic Biothreat, or How We Became Unprepared," *Cultural Anthropology* 23 no. 3 (2008): 399–428; and Masco, *Theater of Operations.*

19. On the use of tabletop exercises in speculative war planning see Matthew Kirschenbaum, "Sand Tables: A Granular History of a Speculative Form," lecture, Center for 21st Century Studies, University of Wisconsin–Milwaukee, April 17, 2015; Andrew Lakoff, *Unprepared: Global Health in the Time of Emergency* (Durham, N.C.: Duke University Press, 2017), 22–23.

20. Lindsey Thomas analyses the ways that scientific policy discourse around preparedness draws from techniques of narrative fiction, thereby collapsing the distinction between the empirical and speculative. Thomas, *Training for Catastrophe,* 58–60.

21. The phrase is Herman Kahn's, a military strategist and systems theorist employed at RAND beginning in the 1960s.

22. On this point Thomas importantly notes "they do not posit that there is no difference between fiction and reality, but rather that fiction has a reality of its own." Thomas, *Training for Catastrophe,* 54.

23. On the endlessness of this position, see Lakoff, *Unprepared,* 13–34; Thomas, *Training for Catastrophe.*

24. Donald A. Henderson, "Surveillance Systems in Intergovernmental Cooperation," in *Emerging Viruses,* ed. Stephen Morse (New York: Oxford University Press, 1996), 283–89. See also Ben Anderson, "Security and the Future: Anticipating the Event of Terror," *Geoforum* 41 (2010): 227–35.

25. It is crucial to note that Lakoff and Masco differ on whether preparedness could ever be a successful mode of governance. Lakoff views preemption as a faceted collection of techniques applied to the governance of critical infrastructures, with some of its dimensions more helpful than others. Masco sees the project of preemption as inextricably braided into institutionalized cultures of secrecy untenable to a functioning of society. See Andrew Lakoff, "From Disaster to Catastrophe: The Limits of Preparedness," *Understanding Katrina: Perspectives from the Social Sciences,* June 11, 2006, http://understandingkatrina.ssrc.org; Lakoff, "Generic Biothreat"; Lakoff, "Further Reflections on 'Two Regimes of Global Health': On the Elision of Distinctions," *Humanity Journal* (blog), June 9, 2014, http://humanityjournal.org; Masco, *Theater of Operations.*

26. Brian Massumi, "National Enterprise Emergency: Steps Towards an Ecology of Powers," *Theory, Culture and Society* 26. no. 6 (2009): 159.

27. For an illuminating account documenting how the project of pandemic preemption around avian flu *necessarily* operates in relation to the absence of its direct, manifest threat, see Carlo Caduff, *The Pandemic Perhaps: Dramatic Events in a Public Culture of Danger* (Oakland: University of California Press, 2015).

28. Raymond Williams, *Marxism and Literature* (Oxford: Oxford University Press, 1977), 129.

29. On media ecological theories that join the sensorial and visible, see Mathew Fuller, *Media Ecologies: Materialist Energies in Art and Technoculture* (Cambridge, Mass.: MIT Press, 2005).

30. Lisa Parks, "Obscure Objects of Media Studies: Echo, Hotbird and Ikonos," *Mediascape: Journal of Cinema and Media Studies* (2007): 1.

31. "Outbreaks Near Me" is the name of HealthMap's real-time, global outbreak surveillance application available online and as a mobile phone application. It was developed by researchers at Boston Children's Hospital in 2006. The CDC's flu calculator is no longer running. Since its inception and launch in 2010, the software has been developed and applied to other platforms such as FluWorkLoss2.0, and FluSurge2.0. For the history of this software see https://www.cdc.gov (last accessed February 21, 2019).

32. See Richard Grusin, *Premediation: Affect and Mediality after 9/11* (New York: Palgrave MacMillan, 2010).

33. Preparedness and preemptive policies immerse their subjects in fictional worlds and make the fictional a modality of living. They also materialize speculative modes of reasoning. On this, see Thomas, *Training for Catastrophe.*

34. Michael Billig, *Banal Nationalism* (Los Angeles: Sage, 1995).

35. Billig, *Banal Nationalism,* 39.

36. Billig, *Banal Nationalism,* 39.

37. Billig, *Banal Nationalism,* 38–43.

38. This quote is taken from a U.S. CDC tweet posted a few months after the swine flu event of 2009. I initially encountered this message on the CDC's flu information Twitter page under the handle @CDCFlu. Since the tweet has not been archived, this is an approximate date of authorship.

39. Joseph Masco, "Survival Is Your Business: Engineering Affect and Ruins in Nuclear America," *Cultural Anthropology* 23, no. 2 (2008): 378.

40. Brian Massumi, "The Future Birth of the Affective Fact: the Political Ontology of Threat," in *The Affect Study Reader,* ed. Melissa Gregg and Gregory J. Seigworth (Durham, N.C.: Duke University Press, 2010), 52–70.

41. Lauren Berlant, *Cruel Optimism* (Durham, N.C.: Duke University Press, 2011), 1–12, 51–93. For an earlier analysis that reads preemptive preparedness as project that produces crisis ordinariness, see Gloria Chan-Sook Kim, "Pathogenic Nation-Making: Media Ecologies and American Nationhood Under the Shadow of Viral Emergence," *Configurations* 24, no. 4 (2016): 441–70.

42. Berlant, *Cruel Optimism,* 85.

43. "Pandemic Influenza," *CBS News,* November 2009, http://www.cbsnews.com, last accessed June 3, 2012.

44. "Mother of All Avian Flu Clusters in Indonesia," *The Blotter,* May 24, 2006, http://blogs.abcnews.com/theblotter.

45. Jen Lamas, comment on ABC News Facebook post regarding flu related deaths (January 15, 2015, 3:34 p.m.) https://www.facebook.com/abcnews.

46. Lauren Berlant, "Genre Flailing," *Capacious: Journal for Emerging Affect Theory* 1, no. 2 (2018): 157.

47. Berlant, "Genre Flailing."

48. Berlant, "Genre Flailing."

49. Giddens, *Consequences of Modernity,* 21–36.

50. Giddens, *Consequences of Modernity,* 19.

51. Giddens argues that we make emotional leaps of faith, what he calls "bargains with modernity," when we trust in expert systems. We do so to hold our awareness of the risky and unaccountable nature of modern systems at bay. This is necessary, according to Giddens, to prevent one from falling into ontological crisis from moment to moment. When "fateful moments" occur, they reveal the fragile nature of one's trust. Giddens, *Consequences of Modernity,* 92–100.

52. Michael Schudson, *Advertising the Uneasy Persuasion: Its Dubious Impact on American Society* (New York: Basic Books, 1984), 230.

53. Schudson, *Advertising the Uneasy Persuasion,* 232.

54. Berlant, *Cruel Optimism,* 15, 82.

55. Ben Anderson, "Affective Atmospheres," *Emotion, Space and Society* 2, no. 2 (2009): 78.

56. Masco, *Theater of Operations,* 18.

57. Quotations are drawn from the following sources, respectively: David Morens and Anthony Fauci, "Emerging Infectious Diseases in 2012: Twenty Years after the IOM Report," *MBio* 3, no. 6 (2012); IBM, *Deadly Migration*; 2014 Lysol advertising slogan, http://www.lysol.ca; Institute of Medicine, *Learning from SARS: Preparing for the Next Disease Outbreak* (Washington, D.C.: National Academies, 2004), 28.

58. On breathing as substance of collectivity, see, for example: Peter Sloterdijk, *Foam, Spheres Volume III: Plural Sphereology,* trans. Wieland

Hoban (Cambridge, Mass.: MIT Press, 2016); Andreas Phillipopoulos-Mihailopoulos, *Spatial Justice: Body, Lawscape, Atmosphere* (New York: Routledge, 2012).

59. "Protection becomes problematic when we realize that not all danger is external—and not simply because, as I've already noted, we cannot always clearly bound our inside from out. The reaction that becomes 'autoimmune' arises from the ever-present chance that we will fight against our own. . . . On militaristic metaphors, the autoimmune danger is multiple: we risk attack from within, and because of this, we are ill-defended against attack from without. The body misrecognizes itself as a stranger, and fails to direct its forces to shoring up its barriers." Karmen MacKendrick, "Vulnerable from Within: Autoimmunity and Bodily Boundaries," *Journal of Somaesthetics* 3, nos. 1 & 2 (2013): 58–67.

60. MacKendrick, "Vulnerable from Within," 62.

61. Benedict Anderson, *Imagined Communities: Reflections on the Origins and Spread of Nationalism* (London: Verso, 1983).

62. See CDC, "The Social Distancing Law Project," http://www2a.cdc.gov; and CDC, "The Social Distancing Law Assessment Template: A Practical Field-Tested Methodology to Assess your Jurisdictions Legal Preparedness for Pandemic Influenza," July 2010, http://www.cdc.gov.

63. Ulrich Beck, *Risk Society: Towards a New Modernity* (London: Sage, 1992), 49.

2. Materializing Emerging Microbes

1. Gina Kolata, *Flu: The Story of the Great Influenza Pandemic of 1918 and the Search for the Virus that Caused It* (New York: Simon & Schuster, 1999), 112–13; David Brown, "Resurrecting the 1918 Virus Took Many Turns," *Washington Post,* October 10, 2005; Michael McKnight, "Into the Wild. Twice. For Mankind," *Sports Illustrated,* May 27, 2020, 259.

2. William Roberts, "Facts and Ideas from Anywhere: The Pandemic of 1918 and the Search for the Killer Virus," *Baylor University Medical Center Proceedings* 32, no. 3 (2019): 468–76, https://doi.org/10.1080/08998280.2019.1619423; Brown, "Resurrecting the 1918 Virus Took Many Turns"; Elizabeth Fernandez, "The Virus Detective: Johan Hultin Has Found Evidence of the 1918 Epidemic that Has Eluded Experts for Decades," *SFGate,* February 12, 2002.

3. Kolata, *Flu,* 262–63; Roberts, "The Pandemic of 1918"; Brown, "Resurrecting the 1918 Virus Took Many Turns"; Ned Rozell, "How an Alaska Village Grave Led to a Spanish Flu Breakthrough," *Anchorage Daily News,* November 23, 2020.

4. Poor weather and shoddy planning delayed their shipment, and the dry ice that he had taken from the campus laboratory to preserve the specimens during the homeward journey had evaporated.

5. Jeffery K. Taubenberger, Ann H. Reid, Amy E. Krafft, and Thomas G. Fanning. "Initial Genetic Characterization of the 1918 'Spanish' Influenza Virus." *Science* 21, no. 275 (1997): 1793–96, https://doi.org/10.1126/science.275.5307.1793. It should be noted that AFIP went on a similar expedition for the same purposes in 1997 in Nome, Alaska. There were only skeleton remains at that site, and no soft tissue from which to extract viral material.

6. Taubenberger et al., "Initial Genetic Characterization of the 1918 'Spanish' Influenza Virus."

7. Taubenberger et al., "Initial Genetic Characterization of the 1918 'Spanish' Influenza Virus"; Douglas Jordan, "The Deadliest Flu: The Complete Story of the Discovery and Reconstruction of the 1918 Flu Virus," Centers for Disease Control and Prevention, December 17, 2019.

8. At that time, medical geographer Kirsty Duncan was getting a search together to exhume the 1918 virus from permafrost. AFIP's team was initially working with her. There are conflicting accounts of why AFIP stopped working with Duncan. Two different accounts can be found in the following sources: Kolata, *Flu*; Kirsty Duncan, *Hunting the Flu: One Scientist's Search for a Killer Virus* (Toronto: University of Toronto Press, 2003).

9. Hultin used his own funds in order to bypass the bureaucratic procedures entailed in government research. Kolata, *Flu,* 258–59.

10. The fat acted as a preservative for the lungs, which would have otherwise become desiccated over the decades.

11. Jefferey K. Taubenberger, Ann H. Reid, Raina M. Lourens, Ruixue Wang, Guozhong Jin, and Thomas G. Fanning, "Characterization of the 1918 Influenza Virus Polymerase Genes," *Nature* 437 (2005): 889–93.

12. Jordan, "The Deadliest Flu."

13. See, for instance, Nicole Shukin, *Animal Capital: Rendering Life in Biopolitical Times* (Minneapolis: University of Minnesota Press, 2009); Mike Davis, *The Monster at Our Door: The Global Threat of Avian Flu* (New York: Owl Books, 2005); Frédéric Keck, *Avian Reservoirs: Virus Hunters and Birdwatchers in Chinese Sentinel Posts* (Durham, N.C.: Duke University Press, 2020); Neel Ahuja, *Bioinsecurities: Disease Interventions, Empire and the Governance of Species* (Durham, N.C.: Duke University Press, 2016).

14. Taubenberger et al., "Characterization of the 1918 Influenza Virus Polymerase Genes."

15. On the genetic revolution and new investments in seeing life, see Evelyn Fox Keller, *The Century of the Gene* (Cambridge, Mass.: Har-

vard University Press, 2000); Eugene Thacker, *Biomedia* (Minneapolis: University of Minnesota Press, 2004); Stefan Helmreich, *Silicon Second Nature: Culturing Artificial Life in a Digital World* (Berkeley: University of California Press, 1998).

16. Colin Milburn, *Nanovision: Engineering the Future* (Durham, N.C.: Duke University Press, 2008), 74–82.

17. Milburn, *Nanovision,* 85.

18. Lily Kay, *Who Wrote the Book of Life: A History of the Genetic Code* (Stanford, Calif.: Stanford University Press, 2000), 175.

19. Kay, *Who Wrote the Book of Life.* See also Keller, *Century of the Gene.*

20. On the virtualization of life in genetic era see N. Katherine Hayles, *How We Became Posthuman: Virtual Bodies in Cybernetics, Literature, and Informatics* (Chicago: University of Chicago Press, 1999), 13.

21. Kay, *Who Wrote the Book of Life.*

22. In the 1960s, microbiologist Carl Woese produced definitive evidence that proved that genes exchanged matter "horizontally," that is, within the same generation. Proof of horizontal gene transfer disrupted the idea that life evolved only in linear "trees" (as in the "tree of life," which expresses the notion that genes pass matter on vertically, from one generation of genes to the next), pointing instead to a model that, Helmreich states, is best described as a "thicket of life." On Woese's experiments and its implications for the study of genetic evolution, see David Quammen, *The Tangled Tree: A Radical New History of Life* (New York: Simon & Schuster, 2018); Jan Sapp and George E. Fox, "The Singular Quest for a Universal Tree of Life," *Microbiology and Molecular Biology Reviews* 77, no. 4 (2013): 541–50, https://doi.org/10.1128%2FMMBR.00038-13. For an illuminating analysis of the history of horizontal gene transfer and its implications for the history and philosophy of biology, see Hannah Landecker, "Antibiotic Resistance and the Biology of History," *Body & Society* 22, no. 4 (2016): 9–52. For a discussion of contemporaneous gene transfer in processes of symbiogenesis, see Lynn Margulis and Dorian Sagan, *Acquiring Genomes: A Theory of the Origin of Species* (New York: Basic Books, 2002). For a discussion of these texts, see Helmreich, *Alien Ocean: Anthropological Voyages in Microbial Seas* (Berkeley: University of California Press, 2009), 73.

23. Referring to a plasmid isolated from the human pathogen *Corynebacterium striatum,* Landecker discusses the fact that segments of its DNA came from "bacteria of different habitats and geographical origin . . . that last shared a common ancestor about 2 billion years ago." Landecker, "Antibiotic Resistance."

24. To be sure, the rise of post-Newtonian thought did not upend and replace Newtonian theories altogether. Rather, Newtonian law still held

in the world of closed systems. Complexity theory became the model for studying open ones. See N. Katherine Hayles, *Chaos and Order: Complex Dynamics in Literature and Science* (Chicago: University of Chicago Press, 1991), 3–8, 16.

25. The term "chaos" is, as Hayles points out, a slight misnomer: chaos theory does not posit that the world is devoid of order, but rather, a world of indecipherable and indeterminate complexity (which may even give rise to ordered systems). For a discussion on the reasons why the term "chaos" took hold, see Hayles, *Chaos and Order,* 3–4.

26. The interconnection of complexity theory and systems biology can be found in their relation to (and mutual origins in) neo-cybernetic theory. Vladamir Vernadsky's concept of the biosphere posits that biology was the main driver of planetary development, rather than geochemistry, as was previously held. James Lovelock and Lynn Margulis developed this idea in the context of microbial evolution, elaborating it through Gaia Theory. In Margulis's later work with Lovelock and Dorian Sagan, she develops a theory of autopoietic life by arguing that the generation of life inevitably produces cycles of waste and regeneration, a series of cumulative and catastrophic pollution events that constitute points of crisis from which new evolutionary pathways develop. See Vladamir Vernadsky, *The Biosphere* (New York: Springer-Verlag, 1998); Lynn Margulis and Dorian Sagan, *Microcosmos: Four Billion Years of Microbial Evolution* (Berkeley: University of California Press, 1997). For an excellent overview and analysis of these texts in the context of the rise of the biotechnology industry see Melinda Cooper, *Life as Surplus: Biotechnology and Capitalism in the Neoliberal Era* (Seattle: University of Washington Press, 2008). On the embeddedness of systems biology and chaos theory in neo-cybernetics, see Bruce Clarke, *Gaian Systems: Lynn Margulis, Neocybernetics, and the End of the Anthropocene* (Minneapolis: University of Minnesota Press, 2020).

27. Hayles, *Chaos and Order,* 6.

28. This builds on a longstanding trope in Western philosophy concerning the relationship between the material and the immaterial, in which matter blocks the view/access to an organizational structure and essence.

29. This was DARPA's (then ARPA) mission statement during the Eisenhower era. Eisenhower created ARPA after the United States was caught off guard when Sputnik was launched by the Soviet Union. The term "surprise" was applied specifically to technological surprise. However, it mobilized emotional management as well. On this latter point see Joseph Masco, *Theater of Operations: National Security Affect from the Cold War to the War on Terror* (Durham, N.C.: Duke University Press, 2014).

30. Helmreich, *Alien Ocean.*

31. Jessica O'Reilly, *The Technocratic Antarctic: An Ethnography of Scientific Expertise and Environmental Governance* (Ithaca, N.Y.: Cornell University Press, 2017).

32. It is important to note that these values of "habitability," and of "undeveloped," "unworked" territory are deeply rooted in colonialist tropes of Western development and science. I return to the colonial implications of this framework in chapters 3 and 4.

33. Much of this material was drawn from the samples extracted by Hultin. Some of the matter was drawn from older samples preserved in paraffin. See Jeffrey Taubenberger, "The Origin and Virulence of the 1918 'Spanish' Influenza Virus," *Proceedings of the American Philosophical Society* 150, no. 1 (2006): 86–112.

34. Ann H. Reid, Thomas G. Fanning, Johan V. Hultin, and Jeffrey K. Taubenberger, "Origin and Evolution of the 1918 'Spanish' Influenza Virus Hemagglutinin Gene," *PNAS* 96, no. 4 (1999): 1651–56 https://doi .org/10.1073/pnas.96.4.1651.

35. Lisa Messeri, *Placing Outer Space: An Earthly Ethnography of Outer Worlds* (Durham, N.C.: Duke University Press, 2016), 30.

36. Messeri, *Placing Outer Space.*

37. Helmreich, *Alien Ocean,* 4

38. Nicholas B. King, "Security, Disease, Commerce: Ideologies of Postcolonial Global Health," *Social Studies of Science* 32, nos. 5–6 (2002): 763–89.

39. For a longer history and context of the move from "international" to "global" health see Theodore M. Brown, Marcos Cueto, and Elizabeth Fee, "The World Health Organization and the Transition from 'International' to 'Global' Public Health," *American Journal of Public Health* 96, no. 1 (2006): 62–72, https://doi.org/10.2105/AJPH.2004.050831. I note, too, that the invention of the emerging microbes concept was soon followed by all manner of interdisciplinary research teams that examined emergent microbial risk across temporal and species boundaries; emerging microbes across plants, animals, and broader ecosystems (permafrost, climate change, one's health, etc.). I return to these aspects through the book.

40. Jennifer Brower and Peter Chalk, *The Global Threat of New and Reemerging Infectious Diseases: Reconciling U.S. National Security and Public Health Policy* (Santa Monica, Calif.: RAND Corporation, 2003). The report carries forward and consolidates thinking that had taken shape since 1989 through the expansionist process of the first George Bush's New World Order. The latter concept, among other things, placed world governance at the center of international affairs, while positioning the United States at the helm of that effort. The 2003 publication also takes up and continues Clinton's preoccupation with biowarfare and

biosecurity. On this, see Judith Miller, Stephen Engelberg, and William Broad, *Germs: Biological Weapons and America's Secret War* (New York: Simon & Schuster, 2002). Clinton also mounted an interagency working group to address emerging microbes and bioterrorism. See National Security and Technology Council, "Global Microbial Threats in the 1990's: Report of the NSTC Committee on International Science, Engineering and Technology, (CISET) Working Group on Emerging and Re-Emerging Infectious Diseases," National Security and Technology Council Annual Report, PDD-NSTC-7, December 19, 1997.

41. Brower and Chalk, *The Global Threat of New and Reemerging Infectious Diseases,* 2.

42. For an example of political philosophies that shaped the neoconservative view that societies need external enemies, see: Carl Schmidt, *The Concept of the Political* (Chicago: University of Chicago Press, 2016), 38, 43–44; Leo Strauss, *Natural Man and History* (Chicago: University of Chicago Press, 1950).

43. On the cultivation of terror as a means of organizing national life in the United States, see Masco, *Theater of Operations.*

44. Brower and Chalk, *The Global Threat of New and Reemerging Infectious Diseases,* 1.

45. Brower and Chalk, *The Global Threat of New and Reemerging Infectious Diseases,* 2.

46. Brower and Chalk, *The Global Threat of New and Reemerging Infectious Diseases,* xix.

47. Brower and Chalk, *The Global Threat of New and Reemerging Infectious Diseases,* xiii.

48. Brower and Chalk, *The Global Threat of New and Reemerging Infectious Diseases,* 105.

49. David Satcher, "Emerging Infections: Getting Ahead of the Curve," *Emerging Infectious Diseases* 1, no. 1 (1995): 1–6.

50. See Niklas Luhmann, "Describing the Future," in *Observations on Modernity,* translated by William Whobrey (Stanford, Calif.: Stanford University Press, 1988), 63–74.

51. Lorna Weir and Eric Mykhalovskiy, *Global Public Health Vigilance: Creating a World on Alert* (New York: Routledge, 2010).

52. Fluid Dynamics of Disease Transmission Laboratory, *Understanding the Fluid Dynamics of Disease Transmission,* https://lbourouiba.mit.edu, last accessed May 2022; Yang Zhao, Brad Richardson, Eugene Takle, Lilong Chai, David Schmitt, and Hongwei Xin, "Airborne Transmission May Have Played a Role in the Spread of 2015 Highly Pathogenic Avian Influenza Outbreaks in the United States," *Scientific Reports* 9 (2019); Kamran Khan, "BioDiaspora: Evidence-Based Decision Making for Emerging Infectious Global Disease Threats," International Civil Avian

Organization, International Civil Aviation Organization, https://www.icao.int/EURNAT/OtherMeetingsSeminarsandWorkshops/CAPSCA EUR/CAPSCA-EUR03/2-8-CAPSCA_BernJune1913slides.pdf.

53. On atmosphere controlling technologies, see Yuriko Furuhata, *Climatic Media: Transpacific Experiments in Atmospheric Control* (Durham, N.C.: Duke University Press, 2022), 38.

54. Furuhata, *Climatic Media,* 38.

55. See Phillip Adey, "Air's Affinities, Geopolitics, Chemical Affect, and the Force of the Elemental," *Dialogues in Human Geography* 5, no. 1 (2015): 54–75; Peter Sloterdijk, *Terror From the Air,* trans. Amy Patton and Steve Corrocan (Cambridge, Mass.: MIT Press, 2009); Timothy Choy, "Air's Substantiations," in *Ecologies of Comparison: An Ethnography of Endangerment in Hong Kong* (Durham, N.C.: Duke University Press, 2011), 139–68.

56. Melody Jue, *Wild Blue Media: Thinking Through Seawater* (Durham, N.C.: Duke University Press, 2020), 7.

57. Using the example of the scuba diver, Jue theorizes the experience of being immersed in alien elements as dislocating and disorienting, since it involves a displacement of our senses in unfamiliar contexts, mediums, and ways of being

58. Anthony Giddens, *The Consequences of Modernity* (Stanford, Calif.: Stanford University Press, 1990), 19.

59. Robert Foster, *Coca-Globalization: Following Soft Drinks from New York to New Guinea* (New York: Palgrave Macmillan, 2011), 18.

60. Peter Sloterdijck, *Foams: Spheres III,* trans. Wieland Hoban (South Pasadena, Calif.: Semiotext(e), 2016), 60.

61. Celia Lowe, "Viral Clouds: Becoming H5N1 in Indonesia," in "Multispecies Ethnography," special issue, *Cultural Anthropology* 25, no. 4 (2010): 625–49.

62. Heather Paxson, "Post-Pasteurian Cultures: The Politics of Raw-Milk Cheese in the United States," *Cultural Anthropology* 23, no. 1 (2008): 15–47.

63. On different models of public health, see Marcus Cueto, Theodore Brown, and Elizabeth Fee, *The World Health Organization: A History* (Cambridge: Cambridge University Press, 2020).

64. For example, see Bisnhupriya Ghosh, *The Virus Touch: Theorizing Epidemic Media* (Durham, N.C.: Duke University Press, 2022); Nicole Shukin, *Animal Capital: Rendering Life in Biopolitical Times* (Minneapolis: University of Minnesota Press, 2009); Neel Ahuja, *Bioinsecurities: Disease Interventions, Empire, and the Government of Species* (Durham, N.C.: Duke University Press, 2016); Priscilla Wald, *Contagious: Cultures, Carriers, and the Outbreak Narrative* (Durham, N.C.: Duke University Press, 2008); Kirsten Ostherr, *Cinematic Prophylaxis: Globalization and Contagion in*

the Discourse of World Health (Durham, N.C.: Duke University Press, 2005).

65. These are specific examples cited by the authors of the 1992 IOM study. Committee on Emerging Microbial Threats to Health, *Emerging Infections: Microbial Threats to Health in the United States,* ed. Joshua Lederberg, Robert E. Shope, and Stanley C. Oaks Jr. (Washington, D.C.: National Academies Press, 1992), 34–48.

66. Committee on Emerging Microbial Threats to Health, *Emerging Infections,* 47.

67. Committee on Emerging Microbial Threats to Health, *Emerging Infections.* Indeed, the study lists these factors (and not new pathogens) as the causes of "emergence."

68. For the origins of the biosphere concept, see Vladamir Vernadsky, *The Biosphere,* trans. David Langmuir (New York: Copernicus, 1998).

69. Bruno Latour, *The Pasteurization of France,* trans. Alan Sheridan and John Law (Cambridge, Mass.: Harvard University Press, 1988).

70. In a related vein, Andrew Lakoff and Stephen Collier examine how public health techniques in the United States shifted from approaches geared toward population security and "vital systems security." Vital systems security develops imagination-based security techniques to draw futures in to view, and applies its techniques to manage what it views as systemic and worldwide vulnerabilities in critical infrastructures. Stephen Collier and Andrew Lakoff, "Vital Systems Security: Reflexive Biopolitics and the Government of Emergency," *Theory, Culture, Society* 32, no. 2 (2014): 19–51, https://doi.org/10.1177/0263276413510050.

71. Jairus Grove, *War and Geopolitics at the End of the World* (Durham, N.C.: Duke University Press, 2019).

72. Elizabeth De Loughry, "The Myth of Isolates: Ecosystems Ecologies in the Nuclear Pacific," *Cultural Geography* 20, no. 2 (2013): 167–84; Laura Martin, "Proving Grounds: Ecological Fieldwork in the Pacific and the Materialization of Ecosystems," *Environmental History* 23 (2018): 567–92; Eleana J. Kim, "Cold War's Nature: Midcentury American Science and the Ecologization of the Korean Demilitarized Zone (1966–1968)," unpublished manuscript.

73. Kim, "Cold War's Nature."

74. Radioisotopes were therefore later applied to various objects and sites (plants, ponds, deserts) to manifest the boundaries of ecological systems and ecological isolates. Indeed, radioisotopes played a central role in the development of American science across a range of fields, such as molecular biology, cancer research, laser research, biochemistry, and more (indeed, it is this exact application of radioisotopes to molecules that Reid and Taubenberger were using to "tag," and render legible, the genetic matter of the 1918 virus). On the use of radiation as manifesting

technology see Anne Creager, *Life Atomic: A History of Radioisotopes in Science and Medicine* (Chicago: University of Chicago Press, 2013).

75. For an insightful analysis of a parallel logic at work in the context of the ways that the United States proliferates bases in "juridically ambiguous" spaces across the Pacific, see Jodi Kim, *Settler Garrison: Debt Imperialism, Militarism and Transpacific Imaginaries* (Durham, N.C.: Duke University Press, 2022).

3. Flightlines and Sightlines

1. Lyle Fearnley, "Wild Goose Chase: The Displacement of Influenza Research in the Fields of Po'yang Lake, China," *Cultural Anthropology* 30, no. 1 (2015): 12–35; Frédéric Keck, "Sentinels for the Environment: Birdwatchers in Taiwan and Hong Kong," *China Perspectives* 21, no. 2 (2015): 43–54.

2. On the difference between disease phylogeny, ecology, and networks, see Lyle Fearnley, *Virulent Zones: Animal Disease and Global Health in China's Pandemic Epicenter* (Durham, N.C.: Duke University Press, 2020).

3. For a fascinating study on the ways that domestic farmers and wet markets in this region troubled researcher's assumptions that these ecologies could be categorized as either wild or domestic, see Lyle Fearnly, "The Birds of Poyang Lake: Sentinels at the Interface of Wild and Domestic," *limn* 3 (June 2013), www.limn.it.

4. Frédéric Keck, *Avian Reservoirs: Virus Hunters and Birdwatchers in Chinese Sentinel Posts* (Durham, N.C.: Duke University Press, 2020); Fearnley, *Virulent Zones.*

5. Paul Farmer, *AIDS and Accusation: Haiti and the Geographies of Blame* (Chicago: University of Chicago Press, 1992).

6. Fearnley, "Wild Goose Chase"; Keck, *Avian Reservoirs.*

7. Orit Halpern, *Beautiful Data: A History of Vision and Reason Since 1945* (Durham, N.C.: Duke University Press, 2014), 8.

8. On the issue of the use of media technologies addressing phenomenological uncertainty, see, for instance, Jimena Canales, *A Tenth of a Second: A History* (Chicago: University of Chicago Press, 2009); Greg Siegel, *Forensic Media: Reconstructing Accidents in Accelerated Modernity* (Durham, N.C.: Duke University Press).

9. On mechanical objectivity, see Lorraine Daston and Peter Galison, *Objectivity* (New York: ZONE Books, 2007).

10. The status of the index as a guarantor of some truth has long been rightly contested in the histories of science and documentary. For more on this, see, for instance, Bill Nichols, *Speaking Truths with Film: Evidence, Ethics, and Politics in Documentary* (Oakland: University of

California Press, 2016); Lorraine Daston and Peter Galison, *Objectivity* (Cambridge, Mass.: MIT Press, 2007), 21–22.

11. Kirsten Ostherr, *Cinematic Prophylaxis: Globalization and Contagion in the Discourse of World Health* (Durham, N.C.: Duke University Press, 2005), 15.

12. Aden Evans, *Logic of the Digital* (New York: Bloomsbury Academic, 2015), 8.

13. David N. Rodowick, *The Virtual Life of Film* (Cambridge, Mass.: Harvard University Press, 2007); Lev Manovitch, *The Language of New Media* (Cambridge, Mass.: MIT Press, 2001); Jean Baudrillard, *Ecstasy of Communication* (Cambridge, Mass.: MIT Press, 2012).

14. Rodowick, *The Virtual Life of Film,* 104–7.

15. Christelle Gramaglia, "Sentinel Organisms: They Look Out for the Environment," *limn* 6 (2013): 24–29.

16. We might also understand the indexical claims of these animal sentinel media in line with Lorraine Daston and Paul Galison's concept of "mechanical objectivity." Here, mediating devices, such as cameras, became important to the development of scientific knowledge production because their recoding capacities were mechanical, and thus seemingly free of human will and intervention. See Lorraine Daston and Peter Galison, *Objectivity.*

17. Bazin applied this to photographic media on account of the mechanical objectivity he imagined inherent in its processes. In Paul Levinson, *The Soft Edge: A Natural History and Future of the Information Revolution* (New York: Routledge, 1997).

18. See Niklas Luhmann, "Describing the Future," in *Observations on Modernity,* translated by William Whobrey (Stanford, Calif.: Stanford University Press, 1988), 63–74.

19. "IBM Data Anthem," YouTube video, .33, posted by Sophie Spreafico, March 17, 2010.

20. Deborah Cowen, "The Labor of Logistics," in *The Deadly Life of Logistics: Mapping Violence in Global Trade* (Minneapolis: University of Minnesota Press, 2014), 91–127.

21. See Lisa Gitelman and Virginia Jackson, "Introduction," in *Raw Data Is an Oxymoron,* ed. Lisa Gitelman (Cambridge, Mass.: MIT Press, 2013), 1–13; Geoffrey C. Bowker and Susan Leigh Starr, *Sorting Things Out: Classification and Its Consequences* (Cambridge, Mass.; MIT Press, 2000).

22. Bousquet is discussing GIS mapping in particular, yet these insights can be applied to other cartographic techniques developed from the generation of real-time geospatial information, such as GPS. For illuminating analysis of GIS mapping and militarized vision, see: Antoine Bousquet, *The Eye of War: Military Perception from the Telescope to the Drone* (Minneapolis: University of Minnesota Press, 2018), 119–52.

23. In this quote, Alan MacEachren and Menno-Jan Kraak are describing some of the ways that GIS mapping has transformed what maps are made to do and the ways they can be used. Alan M. MacEachren and Menno-Jan Kraak, "Research Challenges in Geovisualization," *Cartography and Geographic Informational Science* 28, no. 1 (2001): 3. Cited in Bousquet, *Eye of War,* 148.

24. See Charles Sanders Peirce, *Collected Writings of Charles Sanders Peirce,* ed. Charles Hartshorn and Paul Weiss (Cambridge, Mass.: Harvard University Press, 1958), cited in Helmreich, *Alien Ocean.*

25. Tess Takahashi, "Data Visualization as Documentary Form: The Murmur of Digital Magnitude," *Discourse* 39, no. 3 (2017): 376–96.

26. This formulation is inspired by Christopher Heuer's analysis of European efforts to navigate the Arctic during the Renaissance era. See Christopher Heuer, *Into the White: The Renaissance Arctic and the End of the Image* (Brooklyn, N.Y.: Zone Books, 2019).

27. Anthony Fauci and Anthony Morens, "Emerging Infectious Diseases in 2012: 20 Years after the Institutes of Medicine Report," *mBio* 3, no. 6 (2012).

28. Rosalind Krauss, "Photography's Discursive Spaces: Landscape/View," *Art Journal* 42, no. 4 (1982): 311–18. Cited in Henrik Gustafsson, "Foresight, Hindsight, and State Secrecy in the American West: The Geopolitical Aesthetics of Trevor Paglen," *Journal of Visual Culture* 12, no. 1 (2013): 148–68.

29. Robert Roos, "H7N9 Mystery: Why Does Age Profile Tilt Order?" Center for Disease Research and Prevention News, April 13, 2013, https://www.cidrap.umn.edu. Cited in Carlo Caduff, *The Pandemic Perhaps: Dramatic Events in a Public Culture of Danger* (Berkeley: University of California Press, 2015), 180.

30. Eugenie Brinkema, *The Forms of the Affects* (Durham, N.C.: Duke University Press 2014), 222.

31. Joseph Masco, *Theater of Operations: National Security Affect from the Cold War to the War on Terror* (Durham, N.C.: Duke University Press, 2014).

32. On the image of time as endless duration, see Brinkema, *Forms of the Affects,* 215–22.

33. Brinkema provides an analysis of the aesthetic form of anxiety by examining depictions of open water in Chris Kentis's 2003 film *Open Water.* In it she examines how the film's shot of the ocean (in which two of the film's protagonists float and wait after falling off a tour boat) as half sky and half water creates a line of separation in the film frame. The interruption of that horizontal line by a shark fin is what shows the fragility of the frame, line, horizon, and its incapacity to hold apart two things, to support life above and death below. By interrupting the line,

separation becomes unbounded depth. Here, she argues, is anxiety in aesthetic form: "The puncture of the fin transforms the rectangle at the bottom of the flat projected image into a space marked as having depth, containing a 'below.' That crucial fin, in other words, brings into being the openness of the depths. The break in the line of the sea is the visual analogue to the capacity of material to transgress that separation: that the floor is a surface; that the tension is vulnerable to puncture; that it does not support or hold or ground a permanent separation; that for all the frame marks out proportions of flat expanse, bodies can and will fall under that dividing line that cuts through the image. . . . This line that is put in place only to be broken is this assemblage of accumulated proximities and forms, not quite line, not quite surface, but hovering amorphously in the space that simultaneously marks out and defines." Brinkema, *Forms of the Affects,* 227.

34. U.S. National Commission on Terrorists Attacks, *The 9/11 Commission Report, Final Report of the National Commission on Terrorist Attacks Upon the United States* (Washington, D.C.: U.S. Government Printing Office, 2004), 339–44. Marieke De Goede also examines the institutionalization and weaponization of imagination in the 9/11 Report: Marieke de Goede, "Premediation and the Post-9/11 Security Imagination," *Security Dialogue* 39, no. 2 (2008): 155–76.

35. Masco, *Theater of Operations,* 24; Gustafsson, "Foresight, Hindsight, and State Secrecy."

36. De Goede, "Premediation and the Post-9/11 Security Imagination"; Gustafsson, "Foresight, Hindsight, and State Secrecy."

37. Training and Doctrine Command (TRADOC G-2), "Operational Environments to 2028: The Strategic Environment of Unified Land Operations," August 2012, https://www.benning.army.mil.

38. Thinking around states and borders also came to shift in the context of the sciences of complexity. Once regarded as solid and permanent, under complexity science, many came to see states themselves in ineluctable, systemic relation to this unfixed and indeterminate world, and therefore also given to the fluid and unpredictable conditions of a world made up of complex systems. For some states, these inclement systems were treated as dynamics to be operationalized. Pierre Bélanger, "Live, Living, Lived: A Manifesto for Life," in *Going Live: From States to Systems,* Pamphlet Architecture 35 (New York: Princeton Architectural Press, 2015), 4–12.

39. James Corner, "Not Unlike Life Itself: Landscape Strategy Now," Cited in Bélanger, *Going Live,* 77.

40. On the creative properties of catastrophic life systems see Bélanger, *Going Live*; Melinda Cooper, *Life as Surplus: Biotechnology and*

Capitalism in the Neoliberal Era (Seattle: University of Washington Press, 2008).

41. "We know that the current and future strategic environment will be characterized by uncertainty, complexity and increasingly nuanced relationships. The conditions of the strategic environment must be understood, captured and factored into Army decision-making. . . . The strategic environment is defined, in the context of this estimate, as the set of global conditions, circumstances, and influences that affect the employment of all elements of US national power." TRADOC G-2, "Operational Environments to 2028," https://www.benning.army.mil.

42. Amoore, *Politics of Possibility,* 12.

43. Developing Warren Weaver's thesis that "communication" in the era of cybernetics is "not so much what you *do* say, as to what you *could* say," Halpern argues that in communication theory, information is redefined "not [as] an index of past or present events but as the potential for future actions." Halpern, *Beautiful Data,* 102.

44. Halpern, *Beautiful Data,* 222.

45. Halpern, *Beautiful Data,* 103.

46. In this way, the point of the index in these animal sentinel media is continuous with the risk technologies discussed by Amoore, which seek not to identify probability (techniques such as actuarial science, probability, or auditing that seek to narrow a field of potential to arrive at a greater sense of the probable), but rather according to a "possibilistic" one. The difference between the two, she continues, is that a possibilistic logic "does not deploy a statistical probabilistic calculation in order to *avert* future risks but rather *flourishes in those conditions of declared constant emergency* because decisions are taken on the basis of future possibility, however improbable or unlikely." Amoore, *Politics of Possibility,* 12.

47. Amoore, *Politics of Possibility,* 160–72.

48. Masco, *Theater of Operations.*

49. Andrew Lakoff says that the goal of contemporary risk techniques work, not to prevent bad events, but instead to prepare vulnerable contemporary systems by making them resilient. Halpern and Mitchell develop a similar theory of resilience in their analysis of the "smart mandate," arguing that the telos of smartness is not progress, nor to find solutions to crisis, instead it is to become resilient. Lakoff, *Unprepared,* 38–39; Orit Halpern and Robert Mitchell, *The Smartness Mandate* (Cambridge, Mass.: MIT Press, 2022), 22–26, 168–69. See also Amoore, *Politics of Possibility,* 120–25.

50. Lakoff argues that contemporary systems security is made capable of working from a place of incomplete knowledge precisely because it

develops as a mode of intervention. Halpern and Mitchell argue that resilience is a mode of action premised on the assumption of uncertainty. Lakoff, *Unprepared*; Halpern and Mitchell, *Smartness Mandate*.

51. See Halpern and Mitchell, *Smartness Mandate,* 208–13.

52. See Jennifer Gabrys, *Program Earth: Environmental Sensing Technologies and the Making of a Computational Planet* (Minnesota: University of Minnesota Press, 2016).

53. Richard van Noorden, "Report Disputes Benefit of Stockpiling Tamiflu," *Nature* (April 13, 2014), https://doi.org/10.1038/nature.2014.15022; Sander Herfst, et al., "Airborne Transmission of Influenza A/H5N1 Between Ferrets," *Science* 336, no. 6088 (2016): 1534–41.

4. Fluid Economies of Biosecurity

1. See Nathan Wolfe, *The Viral Storm: The Dawn of a New Pandemic Age* (New York: Henry Holt, 2011), 237.

2. See Brian Walsh, "How to Prepare for a Pandemic," *Time,* May 18, 2009; Wolfe, *The Viral Storm,* 254.

3. Wolfe has often referred to himself in interviews as a "virus hunter." His Twitter handle name is @virushunter. See also Nathan Spector, "The Doomsday Strain: Can Nathan Wolfe Thwart the Next AIDs Before It Begins?" *New Yorker,* December 12, 2010.

4. Wolfe, *The Viral Storm.*

5. Anjali Nayar, "Looking for Trouble," *Nature* 462, no. 10 (December 2009).

6. Wolfe, *The Viral Storm.*

7. In the mid-2010s, GVFI developed a research wing called the Laboratory for Research in Complex Systems (LRC), and a for-profit corporation called Metabiota. As of January 1, 2020, GVFI effectively closed while the LRC and Metabiota became stand-alone bodies.

8. See Paul N. Edwards, "Epilogue: Cyborgs in the World Wide Web," in *The Closed World: Computers and Politics of Discourse in the Cold War* (Cambridge, Mass.: MIT Press, 2011), 353–65.

9. See, for example, Alex Blanchette, *Porkopolis: American Animality, Standardized Life, and the Factory Farm* (Durham, N.C.: Duke University Press, 2020); Nicole Shukin, "Biomobility," in *Animal Capital: Rendering Life in Biopolitical Times* (Minneapolis: University of Minnesota Press, 2009); Andrew Donaldson, "Biosecurity After the Event: Risk Politics and Animal Disease," *Environment & Planning A* 40 (2008): 1552–67.

10. globalviral.org.

11. There is a long racialized and colonial history and politics behind the terms used to organize countries in such terms. Their comparative

nature (for example, "advanced" versus "emerging") sets nations in relations to one another in terms of a false hierarchy in which some lives are accorded higher value than others, while also enabling some groups to exploit and extract from others. In this chapter, I use the language that classifies countries into numbered tiers (as in "first" and "third" worlds), as "developed" and "developing," and as "wealthy"/"rich" versus "low of middle"/"poor" countries. I do so because these are the same terms deployed by the organizations and bodies mapped out in this chapter's analysis of a new twenty-first-century developmental paradigm. My aim is not to naturalize the use of this vocabulary, but the opposite: to help lay bare the perspectives from which such bodies operate. On the history and politics of these naming conventions, see Themrise Khan, Seye Abimbola, Catherine Kyobutungi, and Madhukar Pai, "How We Classify People and Countries—and Why It Matters," *BMJ Global Health* 7, no. 6 (2022), http://dx.doi.org/10.1136/bmjgh-2022-009704; Michael Thompson, Alexander Kentikelenis, and Thomas Stubbs, "Structural Adjustment Programmes Adversely Affect Vulnerable Populations: A Systematic-Narrative Review of Their Effect on Child and Maternal Health," *Public Health Review* 38, no. 13 (2017); Frank Jacobs, "'The West' Is, in Fact, the World's Biggest Gated Community," *BigThink*, October 12, 2019, https://bigthink.com.

12. See, for example, Joseph Masco, *Theater of Operations: National Security Affect from the Cold War to the War on Terror* (Durham, N.C.: Duke University Press, 2014); Lynn C. Klotz and Edward J. Sylvester, *Breeding Bio Insecurity: How US Biodefense Is Exporting Fear, Globalizing Risk, and Making Us All Less Secure* (Chicago: University of Chicago Press, 2009).

13. See, for example, Ursula K. Heise, *Sense of Place and Sense of Planet: The Environmental Imagination of the Global* (Oxford: Oxford University Press, 2008).

14. My account of the relationship between neoliberal economics, innovation and biotechnology, and earth discourse in this section is very much indebted to Melinda Cooper's analysis of this period. Melinda Cooper, *Life as Surplus: Biotechnology and Capitalism in the Neoliberal Era* (Seattle: University of Washington Press, 2008). On the specific point of Fordist production, see 21–24.

15. For a discussion of this report in relationship to futurology and crisis, see Yuriko Furuhata, *Climactic Media: Transpacific Experiments in Atmospheric Control* (Durham, N.C.: Duke University Press, 2022), 59, 122; Cooper, *Life as Surplus.*

16. Donella H. Meadows, Dennis L. Meadows, Jergen Randers, and William W. Behrens III, *The Limits of Growth: A Report for the Club of Rome's Project on the Predicament of Mankind* (London: Pan Books, 1972).

17. Cooper, *Life as Surplus*; Michael Goldman, *Imperial Nature: The World Bank and Struggles for Social Justice in the Age of Globalization* (New Haven, Conn.: Yale University Press, 2005); Jason Moore, *Capitalism in the Web of Life: Ecology and the Accumulation of Capital* (New York: Verso, 2015).

18. Cooper, *Life as Surplus.*

19. David Harvey, *A Brief History of Neoliberalism* (Oxford: Oxford University Press, 2005).

20. Cooper, *Life as Surplus,* 9.

21. See Gopal Balikrishnan, "Speculations on the Stationary State," *New Left Review* 59 (September–October 2009): 5–26; Harvey, *A Brief History of Neoliberalism*; Rebecca Lave, "Neoliberalism and the Production of Environmental Knowledge," *Environment and Society* 3, no. 2 (2012): 19–38.

22. National Science and Technology Policy: Organizations and Priorities Act of 1976, Public Law 94–282, U.S. Congress 6601 (1976), 459–73; For critical discussions on the so-called innovation crisis, see Catherine Waldby and Robert Mitchell, *Tissue Economies: Blood, Organs and Cell Lines in Late Capitalism* (Durham, N.C.: Duke University Press, 2006); Walter Sullivan, "Loss of Innovation in Technology Is Debated," *New York Times,* November 24, 1976; Thomas V. McElhany, "An 'Industrial Innovation Economy' Is Decried at MIT Symposium," *New York Times,* December 10, 1985; Bradley Graham, "Patent Bill Seeks to Bolster Innovation," *Washington Post,* April 8, 1979.

23. National Science and Technology Policy: Organizations and Priorities Act of 1976, section 3a (6).

24. David E. Nye, *American Technological Sublime* (Cambridge, Mass.: MIT Press, 1994), 10.

25. Cooper, *Life as Surplus.*

26. See Vandana Shiva, *Biopiracy: The Plunder of Nature and Knowledge* (Brooklyn, N.Y.: South End Press, 1997); Bronwyn Parry, *Trading the Genome: Investigating the Commodification of Bioinformation* (New York: Columbia University Press, 2004); Waldby and Mitchell, *Tissue Economies*; Cooper, *Life as Surplus.*

27. I am indebted to Melinda Cooper's examination of the relationship between the bioeconomy, resource depletion, and the New Right for the foregoing analysis. Cooper, *Life as Surplus.*

28. Fred Turner, *From Counterculture to Cyberculture: Stewart Brand, the Whole Earth Network, and Rise of Digital Utopianism* (Chicago: University of Chicago Press, 2006).

29. It should also be noted that the U.S. Supreme Court ruling of 1980 in *Diamond vs. Chakrabarty* (a landmark case that determined whether or not living organisms could be patented) was critical to paving the way for the U.S. biotechnology industry. Prior to this case, life forms were

regarded as unpatentable. The court ruled in favor of Ananda Mohan Chakrabarty, who developed a bacterium while working for General Electric. This decision opened up the space for many new forms of ownership for the biotechnology industry by permitting the patenting of genetically modified forms of life. See U.S. patent 4,259,444. For an analysis and critique of the court's decision, see David Drahos and John Braithwaite, *Information Feudalism: Who Owns the Knowledge Economy?* (London: New Press, 2002), 155–59; David Resnik, *Owning the Genome: A Moral Analysis of DNA Patenting* (Albany: State University of New York Press, 2004).

30. This flow of ideas between the economic and the biological/ecological moved in both directions. Prior to the period I discuss here, from the 1930s onward, Joseph Schumpter's economic theories of "creative destruction" (in which economic systems innovate from within through incessant processes of industrial mutation, crisis, and destruction) grew out of theories of evolutionary biology. Theories of complex, crisis-driven transformations were imported from biology into the economic. In turn, molecular biologists frequently referred to evolutionary economics as they reformatted biological models of growth, thus importing a theory of economic systems into biological ones. On the relationship between evolutionary biology and economics, see, for example, Joseph A. Schumpeter, *Capitalism, Socialism, and Democracy* (New York: Harper, 1942); Jean-Baptiste André, Mikael Cozic, Silvia De Monte, Jean Gayon, Philippe Huneman, Johannes Martens, and Bernard Walliser, *From Evolutionary Biology to Economics and Back: Parallels and Crossings Between Economics and Evolution* (Cham, Switzerland: Springer, 2023).

31. Karl Marx, *Capital: A Critique of Political Economy, Vol. 1,* trans. Ben Fowkes (New York: Penguin, 1990).

32. Waldby and Mitchell, *Tissue Economies*; Cooper, *Life as Surplus;* Moore, *Capitalism in the Web of Life.*

33. See Shiva, *Biopiracy*; Cori Hayden, *When Nature Goes Public: The Making and Unmaking of Bioprospecting in Mexico* (Princeton, N.J.: Princeton University Press, 2003).

34. Latour describes how colonial explorers dealt with the problematic of collecting massive or unwieldy natural phenomena or objects back to their respective homes for study. Rather than attempting to bring back with them objects impossible to transport (a river, or a delicate plant, for example), technologies like nibbed pens, and taxidermic tools, enabled the translation of natural phenomena into more transmissible forms (e.g., specimens, skin studies, scientific drawings). Bruno Latour, *Science in Action: How to Follow Scientists and Engineers Through Society* (Cambridge, Mass.: Harvard University Press 1987), 243.

35. For a critique of this fantasy of dematerialization, see, for

example: N. Katherine Hayles, *How We Became Posthuman: Virtual Bodies in Cybernetics, Literature and Informatics* (Chicago: University of Chicago Press, 1999); Parry, *Trading the Genome,* 42–101.

36. Latour, *Science in Action.* See also Eugene Thacker, *Biomedia* (Minneapolis: University of Minnesota Press, 2004); Kashuik Sunder Rajan, *Biocapital: The Constitution of Post-Genomic Life* (Durham, N.C.: Duke University Press, 2006).

37. See also Parry, *Trading the Genome.*

38. Parry, *Trading the Genome,* 71–75, 126–38.

39. See Nye, *American Technological Sublime.*

40. On this see, for example, Shiva, *Biopiracy*; Goldman, *Imperial Nature*; Hayden, *When Nature Goes Public*; Laurelynn Whitt, *Science, Colonialism and Indigenous Knowledge Peoples. The Cultural Politics of Law and Knowledge* (Cambridge: Cambridge University Press, 2009).

41. Goldman, *Imperial Nature.*

42. On benefit-sharing agreements formed around biological extracts, see Hayden, *When Nature Goes Public,* 48–54; Natalie Porter, *Viral Economies: Bird Flu Experiments in Vietnam* (Chicago: Chicago University Press, 2019).

43. World Commission on Environment and Development, *Our Common Future* (Oxford: Oxford University Press, 1987). For a history of the Brundtland Report in the context of the institutional and organizational mandates of the World Bank and the World Health Organization, see Marcus Cueto, Theodore Brown, and Elizabeth Fee, *The World Health Organization: A History* (Cambridge: Cambridge University Press, 2020).

44. The United States was key to the shaping of the 1992 CBD. It ultimately did not sign the treatise because some feared that ratifying the CBD would weaken a nation's sovereign claim over their resources in other territories. It was asserted that the benefit-sharing structures of the CBD were unclear and that avenues to address noncompliant states were weak and unenforceable. For an analysis of these factors and a detailed account of why the United States did not sign the treaty see Hayden, *When Nature Goes Public*; Shiva, *Biopiracy*; William J. Snape III, "Joining the Convention of Biological Diversity: A Legal and Scientific Overview of Why the U.S. Must Wake Up," position paper, Center for Biological Diversity, www.biologicaldiversity.org.

45. For an excellent reading of the ways that biodiversity regulation and governance condition the disappearance of actual biodiversity both in law and in the world, see Anthony Burke, "Bluescreen Biosphere: The Absent Presence of Biodiversity in International Law," *International Political Sociology* 13, no. 3 (2019): 333–51.

46. Shiva, *Biopiracy*; Hayden, *When Nature Goes Public*; Burke, "Bluescreen Biosphere."

47. Hayden, *When Nature Goes Public,* 57.

48. World Resources Institute, World Conservation Union, United Nations Environment Program, Food and Agriculture Organization of the United Nations, and United Nations Educational Scientific and Cultural Organization, *Global Biodiversity Strategy: Guidelines for Actions to Share, Study and Use Earth's Biotic Resources Sustainably and Equitably* (Washington, D.C.: World Resources Institute, 1992), 5. Cited in Hayden, *When Nature Goes Public,* 53.

49. Hayden, *When Nature Goes Public,* 57.

50. Here I am building upon Cori Hayden's analysis of biodiversity as a "storehouse of information not-yet catalogued." Hayden, *When Nature Goes Public,* 53.

51. World Economic Forum, "How Biodiversity Loss Is Hurting Our Ability to Combat Pandemics," March 9, 2020, https://www.weforum.org (accessed March 27, 2021).

52. These terms of exclusion were somewhat vague even prior to the *Diamond v. Chakrabarty* decision. As patent law in such cases centers on human intervention, patent attorneys were successfully claiming the patentability of microorganisms by singling out techniques such as the "modification" of microorganisms by their insertion into a solution or inert matter, or "purification" by removing redundant pieces of genetic material. The emphasis of patent law on invention and manufacture in defining the patentability of microorganisms also leads to another critique of the notion of "global interest": it implies that only countries that have the economic and technological capacity are able to render value from microorganisms, in ways that countries that lack these technologies cannot achieve. On the perilously flexible framing of "human intervention" in biotech, see Drahos and Braithwaite, *Information Feudalism,* 158. On the inequalities inherent in such a system, see Sunder Rajan, *Biocapital.*

53. Human and animal viruses move through two different, but interlinked, systems. On these two systems, see Porter, *Viral Economies,* 151.

54. In 2006, Indonesia stopped sharing virus samples of avian flu/H5N1 with the WHO's global influenza surveillance network that collects, researches, and develops technologies from virus samples to develop tools to address their potential threat. Supari noted that global pharmaceutical companies enjoy untold profits from vaccines and antivirals developed from Indonesian samples, while leaving those most at risk of infection behind. Supari observed that, despite the fact that the H5N1 virus was sourced from Indonesia, Indonesians would continue to be priced out of critical interventions in the event of an outbreak or pandemic, while being barred from accessing any knowledge arising from research, and ineligible for any benefits of developing vaccines and

antivirals from that virus. Pointing to the fact that those most exposed to H5N1 do not have the means to buy drugs needed to fight the virus, she claimed "viral sovereignty" in an effort to redistribute economic advantages arising from this process. India, Brazil, and Nigeria subsequently began to assert claims to viral sovereignty, and the Non-Aligned Movement of 112 developing nations began to follow suit. See Siti Fadilah Supari, *It's Time for the World to Change: In the Spirit of Dignity, Equality, Transparency: Divine Hand behind Avian Influenza* (Jakarta: Sulaksana Watinsa Indonesia, PT, 2008). For analysis of the Indonesian case see Celia Lowe, "Viral Sovereignty: Security and Mistrust as Future Measures of Global Health in the Indonesian H5N1 Influenza Outbreak," *Medical Anthropology Today* 6, no. 3 (2019): 109–32, https://doi.org/10.17157/mat.6.3.662; Stefan Elbe, "Who Owns a Deadly Virus: Viral Sovereignty, Global Health Emergencies, and the Matrix of the International," *International Political Sociology* 16, no. 2 (2022): 1–18. On the rise of claims of viral sovereignty in nonhegemonic countries, see Simone Verzanni, "Preliminary Remarks on the Envisaged World Health Organization's Pandemic Influenza Preparedness Framework for the Sharing of Viruses and Access to Vaccines and Other Benefits," *Journal of World Intellectual Property* 13, no. 6 (2010): 675–96. Cited in Porter, *Viral Economies,* 153.

55. Amy Hinterberger and Natalie Porter, "Genomic and Viral Sovereignty: Tethering the Materials of Global Biomedicine," *Public Culture* 27, no. 2 (2015): 361–86.

56. On the Pandemic Influenza Preparedness Framework, see World Health Organization, "Resolution WHA 64.5: Adoption of Pandemic Influenza Preparedness Framework," May 2011.

57. Porter, *Viral Economies,* 154.

58. Hayden, *When Nature Goes Public,* 53.

59. Paul Hawken, Amory Lovins, and L. Hunter Lovins, *Natural Capitalism: Creating the Next Industrial Revolution* (Boston: Little Brown, 1999).

60. Hawken, Lovins, and Lovins, *Natural Capitalism.*

61. Peter Benson and Stuart Kirsch, "Corporate Oxymorons," *Dialectical Anthropology* 34, no. 1 (2010): 45–48; Robert J. Foster, "Corporate Oxymorons and the Anthropology of Corporations," *Dialectical Anthropology* 34 (2010): 92–102.

62. Hawken, Lovins, and Lovins, *Natural Capitalism,* xi.

63. Hawken, Lovins, and Lovins, *Natural Capitalism,* 87. The authors of this report use this terminology in relationship to green architecture. There, "green both ways" works as a kind of mandate that places a double demand on "green"; that profit is to be gained from investing in nature. I cite it here because it captures the spirit of the entire publication and project more broadly.

64. Benson and Kirsch, "Corporate Oxymorons."

65. See Jason M. Moore, "The Rise of Cheap Nature," in *Anthropocene of Capitalocene: Nature, History, and the Crisis of Capitalism,* ed. Jason M. Moore (Oakland, Calif.: PM Press, 2016), 78–114. See Jason Moore's discussion of the "double internality": Moore, *Capitalism in the Web of Life: Ecology and the Accumulation of Capital* (London: Verso, 2015), 1–8.

66. U.S. Institutes of Medicine, *America's Vital Interest in Global Health: Protecting our People, Enhancing our Economy, and Advancing our National Interests* (Washington, D.C.: National Academies Press, 1997), 11–45.

67. See Lawrence O. Gostin and Rebecca Katz, "The International Health Regulations: The Governing Framework for Global Health Security," *Milbank Quarterly* 94, no. 2 (2016): 264–313, https://doi.org/10.1111%2F1468-0009.12186.

68. See National Science and Technology Council, Committee on International Science, Engineering, and Technology (CISET), *Global Microbial Threats in the 1990s,* especially chapter 5, https://clintonwhitehouse3.archives.gov/WH/EOP/OSTP/CISET/html/toc.html, and IOM, *America's Vital Interest in Global Health,* 31.

69. IOM, *Orphans and Incentives: Developing Technologies to Address Emerging Infections* (Washington, D.C.: National Academies Press, 1997); IOM, *America's Vital Interest in Global Health.*

70. On resilience as global health strategy see Stephen Collier and Andrew Lakoff, "Vital Systems Security: Reflexive Biopolitics and the Government of Emergency," *Theory, Culture, Society* 32 (2015): 19–51.

71. See IOM, *America's Vital Interest in Global Health.*

72. On the history of global health in the context of economic globalization and neoliberalization, and the changing relation between the World Bank and the WHO, see Cueto, Brown, and Fee, *The World Health Organization: A History,* 239–93.

73. This is Gostin's recommendation to the IHR. Lawrence O. Gostin. "The International Health Regulations and Beyond," *The Lancet* 4, no. 10 (2004): 606–7. https://doi.org/10.1016%2FS1473-3099(04)01142-9.

74. IOM, *America's Vital Interest in Global Health,* 31. See also Nicholas King, "Security, Disease, Commerce: Ideologies of Postcolonial Global Health," *Social Studies of Science* 32, no. 5–6 (2002): 763–89, https://doi.org/10.1177/030631270203200507.

75. CISET, *Global Microbial Threats,* especially chapter 3, item 7.

76. See CISET, *Global Microbial Threats*; IOM, *Emerging Infections*; and IOM, *America's Vital Interest.* For the Cold War connection to "epidemiological intelligence," see Elizabeth Fee and Theodore M. Brown, "The Unfulfilled Promise of Public Health: Déjà Vu All Over Again," *Health Affairs* 21, no. 6 (2002): 31–43.

77. The term "small world networks" comes from Lawrence Gostin's recommendations to the IHR. He uses this term in an article critiquing existing IHR measures. His recommendation that developing countries serve as informational resources echoes the same kinds of calls from the IOM, RAND, and other global health institutions. See Gostin, "The International Health Regulations and Beyond."

78. Gostin, "The International Health Regulations and Beyond."

79. Karl Marx, *Grundrisse, Foundations on the Critique of the Political Economy,* trans. Martin Nicolaus (New York: Penguin, 1993).

80. David Harvey, *Marx, Capital and the Madness of Economic Reason* (Oxford: Oxford University Press, 2018), 4.

81. It is worth noting the linkages between this conceptualization of information as an element endowed with incredible powers and the informational dreams of the California counterculturalists I discussed earlier. As part of their one-mind vision, they saw information as the essence of the world, a network of data out of which everything was made and connected. According to this belief, if one could apprehend the world as information alone, it would be possible to access, understand, and govern that realm in ways that transcend the difficulties of terrestrial physicality, politics, and conflict. I understand the production of information along these lines as deeply shaping Wolfe's vision for his pandemic immune system.

82. David Greene, *The Promise of Access: Technology, Inequality and the Political Economy of Hope* (Cambridge, Mass.: MIT Press, 2021).

83. On the economies of biological waste, see Waldby and Mitchell, *Tissue Economies,* 84–110.

84. Edmund Pratt (Pfizer CEO from 1972 to 1991) claimed that Pfizer was losing their market share "because our intellectual property was not being respected in developing countries." Quoted in Michael A. Santoro and Lynne Sharp Paine, *Pfizer Global Protection of Intellectual Property,* Harvard Business School Case 1995, 6. https://hbsp.harvard.edu (accessed July 2019).

85. See Shiva, *Biopiracy*; Hayden, *When Nature Goes Public*; Goldman, *Imperial Nature.*

86. See, for example, Hayden, *When Nature Goes Public*; Goldman, *Imperial Nature.*

87. Drahos and Braithwaite, *Information Feudalism,* xii.

88. Many in the intellectual property industries made such claims. This specific language is that of Barry MacTaggert, chairman of Pfizer International 1981–1991. Barry MacTaggert, "Stealing from the Mind," *New York Times,* July 9, 1982.

89. See Cooper, *Life as Surplus,* 56; Shiva, *Biopiracy*; Drahos and Braithwaite, *Information Feudalism.*

90. Parry, *Trading the Genome*; Cooper, *Life as Surplus*; Waldby and Mitchell, *Tissue Economies.*

91. It is important to note that the definition of microorganisms here become ambiguous. TRIPS article 27 (especially 27.3 and 27.3b) defines microorganisms in expansive terms and discuss the *obligation* to patent them. It has been argued that TRIPS fails to define some of its key terms, for instance "novelty"/"invention," or "microorganism," and this openness has been put to use to the advantage of those claiming intellectual property. On this see Mike Adcock and Margaret Llewelyn, "Microorganisms: Definition and Options Under TRIPS," occasional paper delivered at a discussion meeting of the Quaker United Nations Office, November 23, 2000, https://www.iatp.org; David Resnik, *Owning the Genome: A Moral Analysis of DNA Patenting* (Albany: State University of New York Press, 2004).

92. Hayden, *When Nature Goes Public.*

93. Waldby and Mitchell, *Tissue Economies*; Cooper, *Life As Surplus.*

94. For more on economies of waste in the biotechnology industries, see Waldby and Mitchell, *Tissue Economies.*

95. Michael Flitner, "Biodiversity: Of Local Commons and Global Commodities," in *Privatizing Nature: Political Struggles for the Global Commons,* ed. Michael Goldman (New York: Pluto Press, 1998). Cited in Hayden, *When Nature Goes Public.*

96. On the dematerialization of inventions, see Mario Biagioli, "Intangible Objects: How Patent Law Is Redefining Materiality," in *Objects and Materials: A Routledge Handbook,* ed. Penny Harvey, Eleanor Casella, Gillian Evans, Hannah Knox, Christine McLean, Elizabeth B. Silva, Nicholas Thoburn, and Kath Woodward (New York: Routledge, 2014); Mario Biagioli, "The Dematerialization of Invention," YouTube video, January 27, 2011, https://www.youtube.com.

97. IOM, *Orphans and Incentives,* 25–35.

98. U.S. Congress, *Government Patent Policy: The Ownership of Inventions Resulting from Federally Funded Research and Development: Hearings Before the Subcommittee on Domestic and International Scientific Planning and Analysis of the Committee on Science and Technology, U.S. House of Representatives, Ninety-Fourth Congress, Second Session, September 23, 27, 28, 29; October 1, 1976* (Washington, D.C.: Government Printing Office, 1976), 17. Cited in Waldby and Mitchell, *Tissue Economies,* 146.

99. UNAIDS, *Report on the HIV Crisis,* 1998, https://data.unaids.org.

100. Article 31of the TRIPS agreement states: "Other Use Without Authorization of the Right Holder: Where the law of a Member allows for other use of the subject matter of a patent without the authorization of the right holder, including use by the government or third parties authorized by the government, the following provisions shall be respected: such

use may only be permitted if prior to such use, the proposed user has made efforts to obtain authorization from the right holder on reasonable commercial terms and conditions that such have not been successful within a reasonable period of time. This waiver may be waived by a Member in the case of national emergency or other circumstances of extreme urgency or in cases of public non-commercial use." https://www.wto.org.

101. Parallel importing allows the user to import pharmaceuticals from the cheapest foreign market without authorization of the rights holder. Compulsory licensing allows a country to domestically produce a product without the consent of the rights holder.

102. This was possible because, prior to TRIPS, Brazil and India did not have patents on pharmaceuticals. On this, see Drahos and Braithwaite, *Information Feudalism,* 5–7, 150–68.

103. In 2001 the United States and Europe withdrew from this program for many complicated reasons: the drugs act made people in wealthy countries realize that these drugs and the cost of R&D were extremely and artificially inflated, that R&D and manufacture could be done for far less, and it threatened to draw attention back to the complexities of price-tiering for the U.S. public.

104. On this see Drahos and Braithwaite, *Information Feudalism.*

105. Article 27–1 of TRIPS says that to be considered intellectual property, innovation has to "involve an inventive step, and are capable of industrial application." On the dematerialization of "invention" in the information era, see Drahos and Braithwaite, *Information Feudalism*; Mario Biagioli, "Between Knowledge and Technology: Patenting Methods, Rethinking Materiality," *Anthropological Forum* 22 (2012): 282–99.

106. Cooper, *Life as Surplus,* 56

107. The United States is the largest debtor in the world who manages to evade late and nonpayment punishment because it confers upon itself the rights to defer payment, via regulation.

108. IOM Committee on Emerging Microbial Threats to Health, *Emerging Infections: Microbial Threats to Health in the United States,* ed. Joshua Lederberg, Robert E. Shope, and Stanley C. Oaks Jr. (Washington, D.C.: National Academies Press, 1992).

109. On digital utopianism, see Turner, *From Counterculture to Cyberculture.*

5. Managing the Microbial Frontier

1. Donna Haraway calls this disembodied gaze from nowhere the "god's eye trick." See Donna Haraway, "Situated Knowledges: The Science Question in Feminism and the Privilege of Partial Perspective," *Feminist Studies* 14, no. 3 (2018): 589.

2. In approaching the bioeconomy as a commodity chain my goal is not to trace the "complex connectivities" that come to converge around a commodity as it embeds and is reembedded around the world, nor do I attempt to address the heterogenous nature of commodities as they are taken up and consumed.

3. Brian Larkin, "The Politics and Poetics of Infrastructure," *Annual Review of Anthropology* 43, no.1 (2013): 327–43.

4. Larkin, "The Politics and Poetics of Infrastructure."

5. My description of logistics here draws from a wide range of literature on the topic. See, for example, Jesse Le Cavalier, *The Rule of Logistics: Wal-Mart and the Architecture of Fulfilment* (Minneapolis: University of Minnesota Press, 2016); Deborah Cowen, *The Deadly Life of Logistics; Mapping Violence in Global Trade* (Minneapolis: University of Minnesota Press, 2009); North Atlantic Treaty Organization, "Logistics," https://www.nato.int (accessed December 12, 2021).

6. Paul N. Edwards, *A Vast Machine: Computer Models, Climate Data, and the Politics of Global Warming* (Cambridge, Mass.: MIT Press, 2010).

7. Le Cavalier, *The Rule of Logistics,* 4–7.

8. North Atlantic Treaty Organization, "Logistics."

9. See, for example, Martin Danilyuk, "Capital's Logistical Fix: Accumulation, Globalization, and the Survival of Capitalism," *Environment and Planning D: Society and Space* 36, no. 4 (2017): 630–47; Cowen, *The Deadly Life of Logistics*; Le Cavalier, *The Rule of Logistics*; Levinson, *The Box: How the Shipping Economy Made the World Smaller and the Economy Bigger* (Princeton, N.J.: Princeton University Press, 2016).

10. David Greene, *The Promise of Access: Technology, Inequality and the Political Economy of Hope* (Cambridge, Mass.: MIT Press, 2021).

11. On the movement of biosciences away from the "wet labs" of molecular biology to the "dry labs" of bioinformatics, see Eugene Thacker, *Biomedia* (Minneapolis: University of Minnesota Press, 2004).

12. Levinson, *The Box.*

13. On the processes of making global standards through the International Standards Office, the conditions of membership with that body, and understanding the roles that individual states play in international standard-setting processes, see Walter Mattli and Tim Büthe, "Setting International Standards: Technological Rationality of Primacy of Power?" *World Politics: A Quarterly Journal of International Relations* 56, no. 1 (2003): 1–42.

14. Stephen D. Krasner, "Global Communications and National Power: Life of the Pareto Frontier," *World Politics* 43, no. 3 (1991): 336–66; John J. Mearsheimer, "The False Promise of International Institutions," *International Security* 19, no. 3 (1994–95): 5–49.

15. Mattli and Büthe, "Setting International Standards."

16. Keller Easterling, *Extrastatecraft: The Power of Infrastructure Space* (London: Verso, 2014), 207.

17. See Wolfe, *The Viral Storm*; Nathan Wolfe, "The Jungle Search for Viruses," video, TED, February 2009, www.ted.com.

18. Edwards tracks the development of "the climate" as an object of global governance. The World Meteorological Association was first guided by principles of "voluntary internationalism" in which nation-states share weather data to form a composite picture of the earth's climate. Cold War fantasies of total computation in the 1950s drew the association toward concerted efforts to extend and link existing data networks together: the world would then be infrastructurally connected in a project to collect, process, and generate climate data. Climate data could only come into being once climate infrastructure became globalized. And this was possible only when state laboratories were made compatible with the standards embodied in that infrastructure. Edwards, *A Vast Machine,* 107–227.

19. Paul Edwards, "Meteorology as Global Infrastructure," *Osiris* 21 no. 1 (2006): 230.

20. On infrastructure as a site of cultural and political negotiation, see Lisa Parks, *Cultures in Orbit: Satellites and the Televisual* (Durham, N.C.: Duke University Press, 2005); Easterling, *Extrastatecraft*; Nicole Starosielski, *The Undersea Network* (Durham, N.C.: Duke University Press, 2015).

21. On the movement of biosamples as a value-adding process, see Bronwyn Parry, *Trading the Genome: Investigating the Commodification of Bioinformation* (New York: Columbia University Press, 2004); Natalie Porter, *Viral Economies: Bird Flu Experiments in Vietnam* (Chicago: Chicago University Press, 2019).

22. For examples of decolonial definitions of these terms, and ones related to them, see Vandana Shiva, *Monocultures of the Mind: Perspectives on Biodiversity and Biotechnology* (New York: Zed Books, 1993); Robin Horton, "African Traditional Thought and Western Science. Part I. From Tradition to Science," *Africa: Journal of the International African Institute* 37, no. 1 (1967): 50–71.

23. Complicating the notion that capital works through an "annihilation of space by time," Michael Simpson argues that the dynamics of stoppage and holding (he is particularly interested in the geoeconomics of oil storage tank farms) work to annihilate "time by space." See Michael Simpson, "The Annihilation of Time by Space: Pluritemporal Strategies of Capitalist Accumulation," *ENE Nature and Space* 1, no. 2 (2019): 110–28.

24. Charmaine Chua, Martin Danyluk, Deborah Cowen, and Laleh Khalili, "Introduction: Turbulent Circulation: Building a Critical Engagement with Logistics," *Environment and Planning D: Society and Space* 36, no. 4 (2018): 617–29.

25. "The velocity of circulation—the time in which it is accomplished, is a determinant of how many products can be produced in a given period of time, how often capital can be realized in a given period of time, how often it can reproduce and multiply its value." Karl Marx, *Grundrisse* (New York: Vintage Books, 1973).

26. Simpson, "The Annihilation of Time by Space."

27. U.S. Institutes of Medicine (IOM), *America's Vital Interest in Global Health: Protecting our People, Enhancing Our Economy, and Advancing Our National Interests* (Washington, D.C.: National Academies Press, 1997), 35.

28. IOM, *America's Vital Interest,* 31–50.

29. World Bank, *World Development Report 1993: Investing in Health* (New York: Oxford University Press, 1993), 5; Kamran Abassi, *World Bank/World Health Under Fire; The Human Development Network, Safe Motherhood and the World Bank: Lessons from 10 Years of Experience* (Washington, D.C.: World Bank, 1999); IOM, *America's Vital Interest,* 16–18, 31–33; Jennifer Prah-Rueger, "The Changing Role of the World Bank in Global Health," *American Journal of Public Health* 95, no. 1 (2005): 60–70. See also Marcus Cueto, Theodore Brown, and Elizabeth Fee, *The World Health Organization: A History* (Cambridge: Cambridge University Press, 2020).

30. World Bank, *World Development Report 1993,* 169.

31. For a history tracing the organizational interactions between the World Bank and the WHO, see Cueto, Brown, and Fee, *The World Health Organization.*

32. On the juxtapositions and incommensurabilities of militarized and humanitarian visions of global health, see Andrew Lakoff, *Unprepared: Global Health in a Time of Emergency* (Oakland: University of California Press, 2017), 67–94.

33. IOM, *Orphans and Incentives: Developing Technologies to Address Emerging Infections* (Washington, D.C.: National Academies Press, 1997), 34.

34. IOM, *Orphans and Incentives,* 19.

35. IOM, *Orphans and Incentives,* 23.

36. *Price-tiering* refers to the practice of systematically setting higher prices in higher-income countries and lower prices in lower-income countries. The principle is one of establishing correlation between price and income. Developed countries would be charged more for the same product sold for less in developing countries.

37. The contentious piece for many U.S. politicians was the matter of price-tiering.

38. Mercer Managing Consultant, *Report on the U.S. Vaccine Industry,* commissioned by the Department of Health and Human Services, UNICEF (New York: Mercer Management Consulting, 1995); Mercer

Management Consulting, *Summary of UNICEF Study: A Commercial Perspective on Vaccine Supply* (New York: Mercer Management Consulting, 1994).

39. IOM, *America's Vital Interest,* 38.

40. For a discussion on this matter, see IOM, *Orphans and Incentives*; IOM *America's Vital Interest.*

41. IOM, *America's Vital Interest,* 35–39.

42. IOM, *America's Vital Interest,* 36. My italics.

43. IOM, *Orphans and Incentives,* 19, 95.

44. The *Diamond v. Chakrabarty* decision (1980) was also a critical piece of legislation in this regard.

45. See David Dickson, *The New Politics of Science* (Chicago: University of Chicago Press, 1988); Cooper, *Life as Surplus,* 27.

46. For an insightful analysis of this logic, see Jamie Cross and Alice Street, "Anthropology at the Bottom of the Pyramid," *Anthropology Today* 25, no. 4 (2009): 4–9.

47. IOM, *America's Vital Interest,* 45

48. Cross and Street, "Anthropology at the Bottom of the Pyramid."

49. C. K. Prahalad, *Fortune at the Bottom of the Pyramid: Eradicating Poverty Through Profits* (Philadelphia: University of Pennsylvania, 2004).

50. IOM, *America's Vital Interest,* 35.

51. IOM, *America's Vital Interest,* 23, 37

52. This emphasis on global surveillance can be found across the government literature. For specific discussion on this in *America's Vital Interest,* see IOM, *America's Vital Interest,* 29–33.

53. IOM, *America's Vital Interest,* 35.

54. Melinda Cooper, *Life as Surplus: Biotechnology and Capitalism in the Neoliberal Era* (Seattle: University of Washington Press, 2008), 61–62.

55. See Cooper, *Life as Surplus*; Jason M. Moore, "The Rise of Cheap Nature," in *Anthropocene of Capitalocene: Nature, History, and the Crisis of Capitalism,* ed. Jason M. Moore (Oakland, Calif.: PM Press, 2016), 78–114.

56. Cooper, *Life as Surplus,* 62.

57. See, for example, Daniel E. Slotnik, "'A Scandal': The W.H.O. Says the World's Rate of Booster Shots Outstrips Poorer Countries' Vaccinations," *New York Times,* November 12, 2021; Alexandra E. Petri, "U.S. Ramps Up COVID Boosters as Poor Nations Await First," *New York Times,* October 22, 2021; Agence France-Presse, "WHO Condemns Rush by Wealthy Nations to Give Covid Vaccine Booster," *The Guardian,* August 18, 2020; Peter S. Goodman. "Poorer Nations at Back of Line for the Vaccine," *New York Times,* December 26, 2020.

58. See, for example, Stephanie Nolan, "Here's Why Developing Countries Can Make mRNA Covid Vaccines," *New York Times,* October 22, 2021; Amy Maxman, "The Fight to Manufacture COVID Vaccines in

Lower-Income Countries," *Nature* 597, no. 7877 (2021): 455–57; "Experts Identify Over 100 Plus Firms to Make COVID mRNA Vaccines," Human Rights Watch, December 28, 2021, https://www.hrw.org.

59. World Trade Organization, "Waiver from Certain Provisions of the TRIPS Agreement for the Prevention, Containment and Treatment of COVID-19," IP/C/W669/Rev. 1 (21-4307), May 25, 2021, https://docs.wto.org.

60. IP law surrounds all manner of components in the Covid-19-related supply chain, ventilation valves, testing equipment and treatments, equipment screws and connection joints. Mark Stoller, researcher and writer for the American Economic Liberties Project, points, for example, to the thousands of patents and IP protections surrounding the plastic bags needed for vaccine development, "Single Use Bioprocessing Equipment is the category name for the set of tools that are helpful in biopharmaceutical manufacturing . . . in order to make certain types of medicine, you can either use big metal containers that must be cleaned each time, or plastic bioreactor bags that can be thrown away. . . . SUB equipment allows continuous manufacturing and flexibility in production. SUB isn't just plastic bags, but includes polymer-based disposable bioreactors, containers, purification columns," and more all designed to meet the proprietary specifications through which each component of the SUB is made interoperable. David Barclay and Mark Stoller, "Why Are There Shortages of Plastic Bags Needed for Vaccine Production? Monopolies and Patents," BIG, May 11, 2021, https://www.thebignewsletter.com.

61. PhRMA, "PhRMA Statement on WTO TRIPS Intellectual Property Waiver," press release, www.catalyst.phrma.org, May 5, 2020; Cynthia Hicks, "New Polling Shows Americans Are Sounding the Alarm Bell on the TRIPS Waiver," PhRma, May 14, 2021, https://catalyst.phrma.org.

62. World Trade Organization, "Draft General Council Declaration on the TRIPS Agreement and Public Health in the Circumstances of a Pandemic. Communication from the European Union to the Council for TRIPS," IP/C/W681 (21-5034), June 18, 2021, https://docs.wto.org.

63. Another dimension to consider is that vaccines are often essentially made in two parts: the sourcing of active pharmaceutical ingredients (API), and processing and compounding. Eighty percent of vaccine APIs come from India, China, Taiwan, and parts of Africa. On the inequity of the logistics of vaccine manufacture, see Rebecca Robbins and Benjamin Mueller, "Covid Vaccines Produced in Africa Are Being Exported to Europe," *New York Times,* August 16, 2021; Stephanie Nolen, "Fighting Covid with Vaccines Made Globally," *New York Times,* October 23, 2021.

64. Some have interpreted this waiver for aspects related to compulsory licensing as a delaying and diversion tactic that "slyly" invokes the

"waiver" to repackage existing WTO flexibilities as if they were new, while imposing new limitations and restrictions that would prevent nations seeking to produce vaccines from doing so without violating TRIPS. See, for example, Public Citizen, "EU's 'Waiver' on Some Compulsory Licensing Wouldn't Even Scratch the Surface of the COVID-19 Health Product's Inequitable Access," www.citizen.org; Médecins Sans Frontières, "MSF Comments on the Reported Draft Texts of the TRIPS Waiver Negotiation," March 2022, https://msfaccess.org.

65. See, for example, Sheryl Gay Stolberg, "As Poor Nations Seek Covid Pills, Officials Fear Repeat of AIDS Crisis," *New York Times,* May 8, 2022.

66. Daniel Marens, "New European Vaccine Proposal Offers Limited Help to Developing Countries," *Huffington Post,* October 13, 2021.

67. Tom Cohen, "Introduction: Climate Change and the Defacement of Theory," in *Telemorphosis,* ed. Tom Cohen (Ann Arbor, Mich.: Open Humanities Press, 2012).

68. See Evan Ratliff, "We Can Protect the Economy from Pandemics. Why Didn't We?" *Wired,* June 6, 2020; Alice Park, "Building a Better Vaccine," *Time,* April 15, 2020.

69. Bill Rossi, CEO of Metabiota, states: "Our Pathogen Sentiment Index combines our team's in-depth scientific knowledge with a novel way to measure the potential impact to consumer travel and spending as a result of an outbreak and trigger the necessary resources to offset it. . . . With PathogenRX, we are pushing the boundaries of insurability." Gunther Kraut, head of Epidemic Risk Solutions at Munich Re, adds: "This joint effort is an important part in our continued strategic push for innovation. It will pave a new frontier for the insurance industry by helping businesses become more resolute to the evolving health threats facing the world." Matt Sheenan, "Munich Re and Marsh Collaborate on Pandemic Risk Solution," *Reinsurance News,* May 18, 2018. https://www.reinsurancene.ws. For an incisive reading of computer modeling and predictive technologies in rendering catastrophic risk into an asset form, see Leigh Johnson, "Catastrophe Bonds and Financial Risk: Securing Capital and Rule Through Contingency," *Geoforum* 35, no. 40 (2013): 30–50.

70. Ratliff, "We Can Protect the Economy from Pandemics."

71. On risk and speculative finance, see, for example, Ivan Ascher, *Portfolio Society: On the Capitalist Mode of Prediction* (New York: Zone Books, 2016); Annie McLanahan, *Dead Pledges: Debt, Crisis and Twenty First Century Culture* (Stanford, Calif.: Stanford University Press, 2016); Marieke de Goede, *Virtue, Fortune, and Faith: A Genealogy of Finance* (Minneapolis: University of Minnesota Press, 2005); Leigh Johnson, "Catastrophe Bonds and Financial Risk."

72. Metabiota's work with the insurance and reinsurance sector is a case in point. Metabiota's primary clients are in the insurance sector. They help insurance and reinsurance corporations compose insurance-linked securities. Mapping neatly onto the resolving logic of the global health economy—the "win-win" scenarios mentioned earlier—by mobilizing their predictive techniques, Wolfe hopes to help the world prepare for pandemic risks while also, or by, "helping the insurance industry build new revenue streams at a time when their core business is under threat by new business models and technologies."

73. Thanks to Annie Moore for pointing out that these arrows form a black hole.

74. See Cooper, *Life as Surplus*, 96–99.

75. From a talk to prospective microbiology students at Stanford. A version of that lecture is available at Nathan Wolfe, "What's Left to Explore?" video, TED, February 24, 2012, www.ted.com.

76. Elizabeth Povinelli, *Empire of Love: Towards a Theory of Intimacy, Genealogy, and Carnality* (Durham, N.C.: Duke University Press, 2006), 76–84.

Index

Page numbers in italics refer to figures.

Gloria Chan-Sook Kim is assistant professor of media and culture at the University of California, Riverside.